Practical Cheesemaking

PRACTICAL CHEESEMAKING

KATHY BISS

The Crowood Press

First published in 1988 by
The Crowood Press
Ramsbury, Marlborough,
Wiltshire SN8 2HE

© Kathy Biss 1988

All rights reserved. No part of this publication may be reproduced or transmitted in any form or by any means, electronic or mechanical, including photocopy, recording, or any information storage and retrieval system without permission in writing from the publishers.

British Library Cataloguing in Publication Data
Biss, Kathy
Practical cheesemaking.
1. Cheeses. Making – Manuals
I. Title
637'.33

ISBN 1-85223-023-1

Line drawings by Janet Sparrow

Typeset by Acorn Bookwork, Salisbury, Wilts
Printed in Great Britain by Butler & Tanner Ltd, Frome and London

Contents

	Acknowledgements	6
	Introduction	7
1	Cheesemaking Equipment	13
2	Milk for Cheese	25
3	Starter	31
4	Rennet and Additives	40
5	Hygiene, Water Supply and Effluent	48
6	Control Tests and Cheese Records	53
7	The Cheese Recipe	63
8	Practical Cheesemaking	74
9	Remedies and Alternative Actions	96
10	Presswork and Storage	105
11	Grading and Packaging	116
12	Cheese Recipes	122
	Appendix: Titratable Acidity of Milk	135
	Glossary	136
	Further Reading	139
	Useful Addresses	140
	Index	143

Acknowledgements

I would like to thank everyone who has had some involvement with the production of this book. My appreciation goes particularly to my aunt Miss E. Eales who undertook to type the original draft, my husband David who typed the final copy and to all my family – Christine, Kate, Robert and 'Gran', who helped in some way.

I would also like to acknowledge assistance with the photography from David Lane, Collin Polson, The States of Guernsey Dairy, Somerset College of Agriculture and Horticulture, and R. and G. Wheeler of Dunchideock.

Thanks must also be extended to all the students that I have had contact with over the past fifteen years, whose confusion with the cheesemaking process has influenced my approach to writing this book.

Introduction

Cheesemaking is a most fascinating craft that can be practised by people from all walks of life – it is just as possible to make cheese in a bowl in the kitchen as it is to make it in a purpose-built vat. It is also a science, and in order to make a good cheese of consistent quality, it is important both to understand the science and to learn the craft.

The aim of this book is to provide all the information required to manufacture an acceptable hard cheese. It is intended to enable the prospective cheesemaker to have some understanding of the science involved and to lead him to an appreciation of the craft skills that are needed.

The cheesemaking process is described in practical detail, and information about equipment, milk, bacterial cultures and other ingredients is included. The process is explained simply both in cheesemaking and in scientific terms, and experience gained in teaching the theory and practice of cheesemaking has influenced this choice of approach. Students at all levels seem to find that a simple science-based explanation of apparently complex cheesemaking concepts helps them to form a clearer idea of the 'mechanics' of the process. The manufacture of cheese is complicated, but oversimplified explanations lead to misconceptions and are responsible for some of the 'cheese mythology' that exists.

This book should provide a useful accurate source of the basic techniques needed for all cheesemaking, whether it is done commercially or for fun.

How to Use This Book

In order to achieve the aims set out, the cheesemaker has a considerable task ahead of him. It is important to study the early chapters in this book very carefully, before embarking upon the actual manufacture of the cheese. These chapters provide a great deal of background information and a knowledge of this is essential in order to understand the process fully. The function and action of the various

INTRODUCTION

ingredients used is explained, and information about equipment and other important aspects of cheesemaking is given.

The method of cheese manufacture described in this book applies particularly to Cheddar cheese, and the practicalities involved in the process are explained in detail, always assuming that the cheesemaker has had no previous practical experience.

The first task is to copy out the contents of a recipe sheet (found at the end of the introduction) – this copy should be kept alongside the book while all the various stages of manufacture are studied. This will help the cheesemaker in his interpretation of the process, which has a language of its own. When the cheesemaker feels that he understands the process reasonably well it could be the stage at which to make a trial cheese.

The remaining part of the book will help with assessing the quality of the cheese and, subsequently, its sale. There is also a selection of hard and semi-hard cheese recipes which could be tried when the reader is sufficiently experienced.

The Cheesemaker

Cheesemakers may be people from all walks of life – smallholders, farmers, dairy and food technology students, home economics students, or anyone who would just like to try their hand at making cheese at home.

All cheesemakers must be conscientious, observant and patient, in order to develop the skills needed to make the cheese, and to maintain high standards during the cheesemaking process. Patience is the most tried of these qualities, since, from the time the cheese is made until it is mature, there is a delay of three to six months. This always proves to be a very frustrating period although, hopefully, the result is worth waiting for.

The History of Cheese

Cheesemaking is a fascinating craft which probably originated at the time of prehistoric man. Early nomadic tribes stored surplus milk in pouches made from animal stomachs, and these pouches were carried on the backs of the pack animals as the tribe travelled. The combination of warmth, souring, and enzymes from the stomach lining coagulated the milk, forming a curd. This curd had much better keeping qualities than the milk and was a very primitive form of

INTRODUCTION

cheese. The milk used in those days would have been taken from such animals as sheep, goats, buffaloes and yaks.

The next significant development in cheese history was in 54 BC when the Romans introduced the first form of organised cheesemaking in Britain, during their occupation. They had travelled across Europe before settling in Britain, and during this period they were able to observe many different methods of making cheese. These skills were passed on to the native Britons, and considerable progress was made in those areas of Roman settlement where the land was most suited to grass growing. The milk production was mainly from sheep, and the areas where it took place were Cheshire, Somerset, Derbyshire and Lancashire – the homes of some of the best-known English cheeses.

The next notable advance in cheese history was during the time of William the Conqueror. From the year 1066 many monasteries and abbeys were established and the various orders of monks were responsible for developing some of the traditional English cheeses. The Cistercian monks at Jervaulx Abbey in Yorkshire are probably the best example of this – they developed Wensleydale cheese from ewes' milk.

After the dissolution of the monasteries in 1536, much of the craft was lost until about 1600. Between 1600 and 1800 there was a great deal of interest and progress in the breeding of cows specifically for milk production. The improvement in cattle-breeding techniques led to increased yields of milk and the need to use these greater quantities profitably. Increased milk production also coincided with improvements in the methods of travel and people from the country areas were now able to sell cheese in the larger cities. Somerset farmers were especially ideally situated. The milk was produced on the Somerset Levels, just south of the Mendip hills (an area ideal for growing grass), and the cheese was made in the local farmhouses. It was then stored in the caves at Cheddar. These provided an excellent environment because they were humid, and remained at a constant temperature all year round.

Progress continued gradually during the eighteenth and nineteenth centuries, and by the year 1877 it had become commercially viable to form a co-operative dairy. The first was formed in Derbyshire and factories continued to develop in Britain up until 1950. At this time, mechanisation began to take over from the more traditional systems which had been used, and smaller factories were closed down.

Despite mechanisation, cheesemaking remains a craft, and will do so until the characteristics and behaviour of the micro-organisms used can be controlled.

INTRODUCTION

Definition and Classification

It is difficult to define cheese nowadays because there are so many ways of manufacturing milk into a curd form. The Food and Agricultural Organisation have, however, given the following definition in a 'Code of Principles'. It does not apply to all cheeses but provides a good general description.

'Cheese is the fresh or matured product obtained by the drainage (of liquid) after coagulation of milk, cream, skimmed or partly-skimmed milk, butter milk or a combination thereof.'

It is just as tricky to classify a cheese as it is to define it. The simple way is to use groupings such as hard, semi-hard, blue-veined or soft.

The hard cheeses are those which have to be ripened for three to twelve months. They have a moisture content of between 38 and 40 per cent, and they include cheeses such as Cheddar, Cheshire, Derby and Double Gloucester.

The semi-hard cheeses have a slightly higher percentage of moisture and the ripening time of the finished cheese can be between ten days and three months. The cheeses in this category are Wensleydale, Caerphilly, White Stilton and Lancashire.

The blue-veined cheeses are generally semi-hard cheeses in which mould growth has been encouraged by the addition of a mould culture. The method of manufacture is designed to encourage this growth during storage. Stilton, Wensleydale, Cheshire and Dorset Blue are examples of blue-veined cheeses made in Britain.

Soft cheeses have a high moisture content and a short period of ripening, which varies according to the type of cheese being manufactured. The main examples of these are Coulumier, Cambridge and Colwick. Mould-ripened varieties have a much longer ripening period and these include Brie, Pont l'Evêque and Camembert.

The Cheesemaking Process

The cheesemaking process is long and complex, and it is necessary to define its stages, and the various ingredients used, this early in the book. This will enable the cheesemaker to gain the greatest benefit from his study of the first few chapters.

A cheese recipe is a guide to the ingredients needed in the process, and also gives details of the operations to be carried out. The recipe will describe the ingredients, the operations, the acidities, and the temperatures at each stage of the process.

INTRODUCTION

Contents of a Cheese Recipe

Ingredients

1. *Milk*   Quality and type.
2. *Starter*   Type, usage rate, temperature and growth rate.
3. *Colour*   Amount as demanded by the variety of cheese or the composition of the milk.
4. *Coagulants (Rennets)*   Rate of usage, acidity of the milk, temperature and setting period.
5. *Salt*   Rate or time of application; dry salting or brining.

Treatment of Ingredients

1. *Milk treatment*   Cooling after production or heat treatment for the destruction of pathogenic bacteria and other unwanted microorganisms.
2. *Starter*   Temperatures, ripening times and acidities.
3. *Coagulation*   The conditions for forming a curd with the milk. These will include acidity, temperature, and method of application.
4. *Cutting*   The curd is cut to release the whey. The size of curd particles along with the rate of scalding affects acid development and cheese moisture.
5. *Scalding or cooking the curd*   The degree and rate of cooking affects the rate of acid production and the rate of curd shrinkage. The temperature is raised very gradually over a period of time.
6. *Stirring*   After cooking the curds are stirred in the whey and further shrinkage of the curd occurs.
7. *Pitching*   The curds are allowed to settle when fit.
8. *Whey off*   The whey is removed when the desired acidity has been reached.
9. *Cheddaring or texturing*   The curd is blocked and turned, in accordance with the variety, in order to gain the texture required.
10. *Milling*   The curd is broken or cut into suitable-sized pieces, ready to be salted.
11. *Salting*   Methods of salting, brining or dry salting of the curd and time of application.
12. *Moulding, vatting or hooping*   The curd is placed in cheese moulds, containers specifically designed to shape the cheese.
13. *Pressing*   The cheese is pressed in the moulds to assist the final removal of the whey. The pressure and rate of application of pressure are defined.

14. *Finishing*   Press work, the final treatment of the cheese before storage.

15. *Storage, maturing*   The temperatures, humidity and conditions of storage and the approximate storage times.

1
Cheesemaking Equipment

Great care must be taken when choosing equipment for use with milk or milk products because milk is an ideal medium for the growth of bacteria (especially when it is warm), and it will easily turn sour and produce 'off' flavours. All equipment which is likely to come in contact with milk must have a smooth surface which can be easily cleaned and sterilised, and which will not impart any taint to the milk.

It must be remembered that during the cheesemaking process lactic acid is produced, and this will have a corroding effect on some surfaces. An example of this is found with aluminium equipment. If aluminium utensils are used it is advisable to try to replace them with glass or stainless steel as soon as it is financially possible.

All equipment used, up to the stage of putting the curd into the moulds, must be easy to clean and sterilise and not subject to corrosion. Once the curd is in the moulds and ready to press, it should be covered by a cloth to protect it from contamination. It is possible at this stage of the process to use wooden followers and boards.

Basic Equipment

Buckets

Stainless steel buckets are recommended for use with milk. Plastic buckets are good if they are kept strictly for milk and are not used for other purposes.

Boards

There are very good plastic boards available now and these are easy to clean. Wooden boards are suitable if kept scrupulously clean and stored in a dry place when not in use.

Cloths

Cloths will be necessary for several purposes during the cheesemaking process. At the beginning of the process the cheese must be kept warm during ripening and renneting, so a hemmed cotton sheet of appropriate size is useful to cover the vat. A similar hemmed sheet would be needed to prevent the curd from becoming chilled whilst cheddaring. This is particularly important in the winter months.

Cloths are mainly required for lining the cheese moulds ready to receive the cheese curd. Traditionally these are cheese-grey (*see* Glossary) on the first day of pressing and smooth calico on the second day. They are available from smallholder suppliers and should be boiled before use.

If a cheesemaking factory is situated in the vicinity it may be possible to purchase disposable nylon net cloths. These are very useful for small cheeses because they do not stick, and can be used instead of the traditional cheese-grey and calico.

The cloth for bandaging the cheese is normally a lightweight calico. For those with economy in mind, old cotton sheets (well boiled) make ideal bandaging material. Thin muslin cloths should not be used. The purpose of the cloth bandage is to give the cheese protection and to prevent it from drying out during storage. The very thin bandage material allows too much air through when the traditional system of bandaging is used.

Thermometer

A floating dairy thermometer is essential, preferably with a clear plastic cover. It can either be in the Celsius or Fahrenheit scale and it is advisable to check its accuracy regularly. The local Trading Standards Office or a hospital laboratory should have an accurate thermometer which might be used for this purpose. Store thermometers carefully to avoid breakage.

Milk Cans

The cans are normally two, five or ten gallon capacity and are most commonly made of aluminium. Stainless steel cans are better, but are expensive. The ten gallon size can often be found redundant on farms – it is also possible that a dairy might have a few that the owner would be willing to sell. The smaller sizes are available from smallholder suppliers.

A small 2–4 pint (1–2 litre) can will be necessary for the production of starter. Keep the cans clean and dry at all times during storage – cans stored in a wet condition are a serious source of contamination.

Sundries

Items such as sieves, measuring jugs, curd brushes and curd knives come into this category. Sieves and measuring jugs can be metal or plastic and of the type normally purchased for the kitchen. A curd brush must be plastic with plastic bristles and a short handle and it is advisable to mark it clearly 'curd'. Curd knives are used to cut the curd into blocks after the whey has drained off. They must be sharp and strong with round-ended blades.

Cheese Mill

This a useful item of equipment, but is not absolutely essential for small-scale production. It works by using a series of fingers or blades to chop the cheese curd into uniformly-sized pieces ready for salting. It is difficult to find a small mill to buy new or second-hand – small mills are sometimes listed in farm auctions but are sold at a very high price. It is possible to mill the cheese using sharp knives, although with quantities of 20 gallons (90 litres) and above it would be very time-consuming. If a mill is found to be essential perhaps a local engineer or blacksmith could make one. A museum with a dairy collection may have one which could be sketched or photographed (with the permission of the curator) and copied.

Small-Scale Production

Jacketed Vat

This is an expensive item of equipment. The vat should be made of stainless steel or tinned steel and be surrounded by a hollow jacket, normally made of galvanised steel. The jacket contains a water or steam heating system. The vat has a tap at the base of one end in order to drain off the whey, and the jacket has a small drainage tap to allow the water or steam condensate to escape.

The jacket can cover the sides and base of the vat, or it may cover only part of it. The former is a better design because the milk and curd can be kept at a more consistent temperature and do not become chilled during manufacture.

The vat illustrated in Fig 2 is heated by pouring hot water into the

Fig 1   A water jacketed vat of up to forty-five litres capacity. The vat has a main milk tank with a large tap to draw off the whey; it also has a water jacket covering one end and the underneath of the vat. Hot water is poured into the jacket via a hinged flap on the top at one end and is drawn off by means of a small tap at the base at the other end.

jacket using the flap at the back. Excess water is drained off through the drainage tap at the front.

When purchasing a vat there are several points to bear in mind:

1. The capacity of the vat. Take into account any increases in production that are likely – in the long term.
2. The design and quality of the vat. Vats suffer a great deal of wear, so it is important that it should be strongly made. Check that all welds and seams are smooth and well finished – they could be serious sources of contamination if milk residue is trapped in them.
3. The availability of spare parts. These should be easily and quickly obtainable.

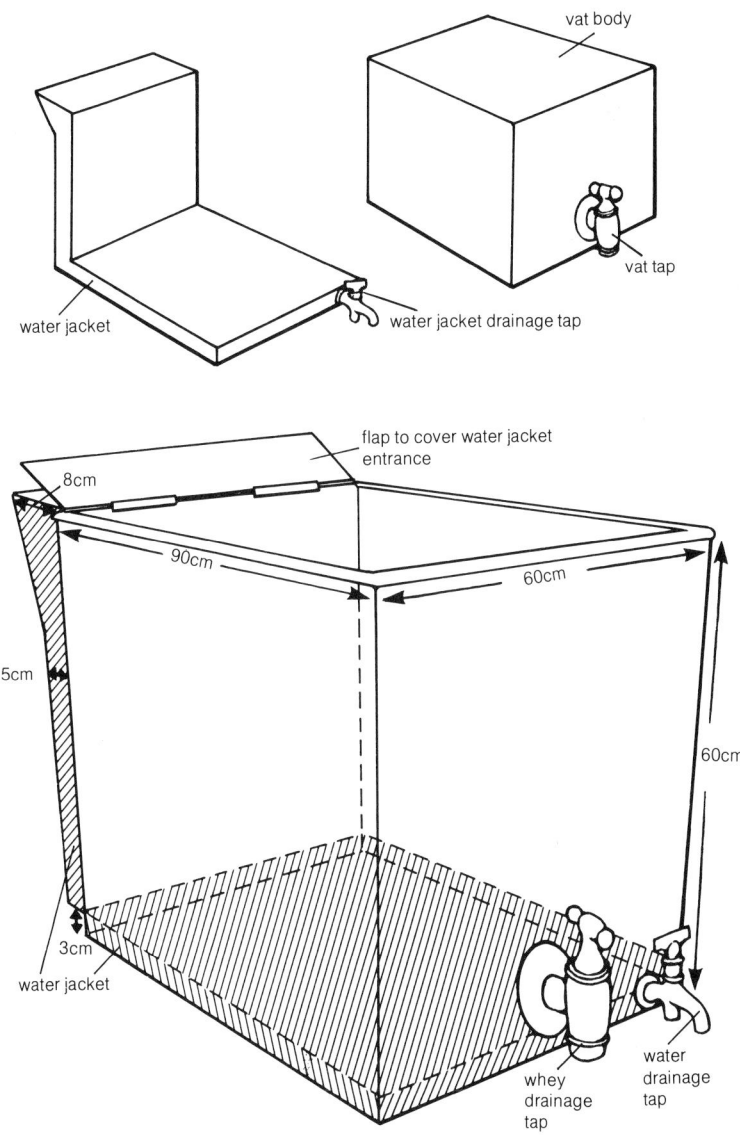

Fig 2 A small forty-five litre vat designed originally for travelling around Somerset schools to demonstrate cheesemaking. The vat has a main tank with a whey drainage tap, and a water jacket with a flap entrance and a tap for drainage.

American Curd Knives

These are essential for cutting the curd into evenly sized pieces. There are several types available, but those most frequently used are knives which carry the blades horizontally or vertically – one of each type will be required. The knife in Fig 36 is a very old design and the blades are positioned diagonally – this eliminates the need for two knives. To cut the curd the knife is reversed at intervals during cutting. The knives are normally made of stainless steel or tinned steel and are often strung with cheese wire instead of blades, with the advantage that it is easily replaced.

Plunger

This is a useful tool which is made from stainless steel. It is used for mixing milk in cans or for stirring in the vat. The disc at the base may be stainless steel or plastic, and the stem and handle are of stainless steel.

Cheese Moulds and Followers

Cheese moulds are available in many shapes and sizes, and are usually made of tinned steel, stainless steel, plastic or wood. They are containers, normally cylindrical in shape, that are designed to take the salted curd to enable it to be pressed. Moulds for hard pressed cheese, which is subjected to high pressures, must be reinforced at the top and base.

Followers are the discs of plastic or wood which are positioned on top of the cheese prior to pressing. As the pressure is applied, the followers sink into the mould on top of the cheese and the curd is thus compressed.

All cheese moulds and followers must be strong, easy to clean and resistant to corrosion from the salt, fat and lactic acid which is found in the finished cheese.

There are many sizes of moulds available. The smallest, or 'miniature' sizes hold 1–4kg (2–9lb) – these are normally hollow tubes with drainage holes. The truckle size of 4–5kg (9–11lb) is also a hollow tube, but has a removable base and reinforcing rings around the top and bottom edge. Larger cheeses can be made in the traditional round moulds or block moulds. These moulds normally take a set weight of 18kg (approximately 40lb) of cheese and they are constructed of aluminium, tinned steel or stainless steel.

Fig 3   A traditional truckle mould for hard pressed cheese with beech wood followers. The mould is of tinned steel with reinforced rims around the top and the base. The cloth lining the mould is a disposable nylon net.

Cheese Press

The press must be able to exert sufficient pressure to force the whey to be dispelled and the curd to be compressed. For a cheddar truckle this amounts to $120kN/m^2$ (20cwt).

There are many presses on the market with various attributes. The most useful press for small producers and home cheesemakers is the Wheeler press, which is very reliable and suitable for small cheeses of up to truckle size of 4-5kg (9-11lb). It uses a system of springs in order to exert pressure on the cheese curd, is easily cleaned and can be dismantled when not in use.

It is still possible to purchase the traditional lever presses, and, although many are antique, they are still efficient and will press several cheeses at a time. The advantage of this press is that it will take cheese of up to 18kg (40lb).

When buying a press try to see as many types as possible in action before deciding which will be the most suitable for your system.

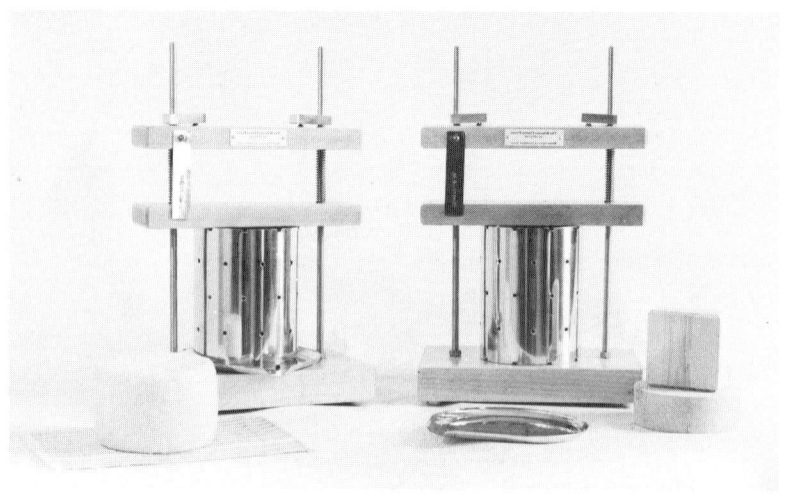

Fig 4 A Wheeler press suitable for cheese of up to 5kg (11lb). The press is appropriate for use with small or larger quantities of curd.

Fig 5 A cheese press made by a dairy engineer in Somerset, showing clearly the main screw and lever mechanism. The weights can be adjusted according to the type of cheese being made.

CHEESEMAKING EQUIPMENT

How a Lever Press Works Cheese presses combine a compound lever system with a screw mechanism and Fig 6 illustrates the arrangement of the former.

A_1 is the extremity of the lower lever that is fixed rigidly to the framework of the press acting as fulcrum at C_1. D_1 is the threaded nut through which S, the main screw of the press, passes; it is also the point at which the upthrust from the cheese is exerted. B_1 is the other extremity of the lower beam, and this is linked to D_2, a point on the upper lever.

A_2 is the extremity of the upper beam which is fixed rigidly to the framework of the press, acting as fulcrum at C_2; B_2 is the other end of the upper beam from which weights are suspended by means of a pulley P and chain C. One end of the chain carries the weights W, the other end of it is fixed on to the press framework.

When the main screw has been turned so that the plunger bears against the cheese, it can no longer move downwards. If one now continues to force the screw to turn further by means of the capstan, the threaded nut D will tend to rise, owing to the screw mechanism. With the exertion of sufficient force on the capstan, the threaded nut D will lift the lower lever, which in its turn will lift the upper lever – this will cause the pulley P to rise and lift the weight W from its shelf.

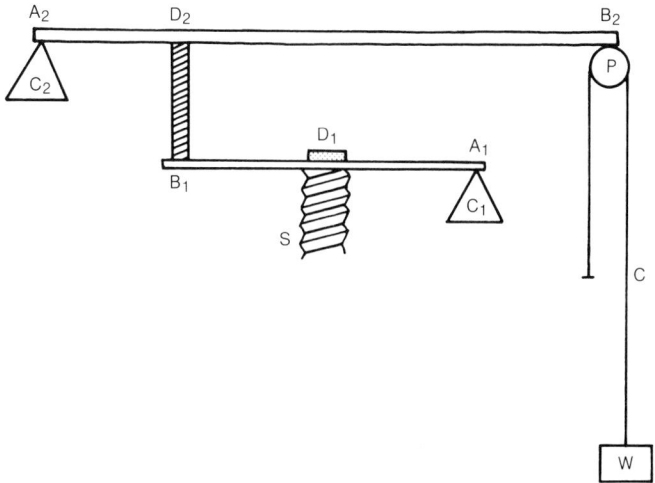

Fig 6   A cheese press in diagrammatic form.

Acidimeter

This piece of apparatus is used to determine the acidity of the cheese at the various stages throughout the cheese make. It consists of a squeeze-fill burette, a white china bowl and glass stirring rod, a 10ml pipette and a dropping bottle. The reagents needed are N/9 Sodium Hydroxide and Phenolphthalein.

If cheese is made regularly, this equipment is essential to provide an accurate measurement of the progress of acidity during each day's make.

It will be necessary to provide a small table or shelf in the cheeseroom to house this equipment, in order to protect it from damage and to keep it readily available for use. The acidity test results must be recorded, so a record sheet and pencil should be kept alongside the equipment.

Kitchen Cheesemaking

The Vat

A large mixing bowl or stainless steel pan is required, to act as a vat. In order to heat this bowl or pan, it will be necessary to place it on a rack, or in a larger bowl of water. An old-fashioned porringer of suitable size makes a good cheese vat.

Cheese Knives

A metal spatula, or a long-bladed knife with a rounded end to the blade, will make a good cheese knife. It is essential that the blade is long enough to reach the bottom of the curd, because it is important that the handle does not touch the curd surface – this will cause damage.

Stirring

This can be carried out using an assortment of spatulas or spoons. It should be borne in mind that the utensils must not impart taints to the cheese, so wooden spoons are not suitable.

Fig 7 For small-scale cheesemaking many utensils found in the kitchen will be suitable. It will be necessary to acquire a thermometer, moulds, followers and mats.

Cheese Moulds and Followers

Small plastic cheese moulds and followers can be purchased from most smallholder suppliers and these are suitable for home and small-scale production. It is difficult to find suitable tins or containers for home-made cheese moulds, as they have to be very strong. It is advisable to purchase purpose-made moulds to prevent a mishap and loss of cheese.

Cheese Press

If a small cheese press is not available use a heavy weight on top of the cheese to compress the curd. A car jack can also be used very effectively. The cheese is placed on the collapsed car jack underneath a strong shelf or beam. The jack is carefully expanded and the cheese is compressed between the jack and the upper surface. This is a very effective press, but the jack will become covered with whey and must be cleaned off promptly before it corrodes. Take care not to apply too much pressure.

It is not possible to measure the degree of pressure when using an

CHEESEMAKING EQUIPMENT

Fig 8 An improvised cheese press can be used successfully. A car jack is positioned above or below the mould and compresses the cheese against an overhead beam.

improvised press, but with experience it will be possible to gauge whether the cheese is sufficiently pressed or not.

Acidimeter

An acidimeter is very useful, even on a small scale and especially if used by an experienced cheesemaker. It is recommended in order to produce consistent results. The apparatus can be purchased from smallholder suppliers, and the reagents can be supplied by the local chemist, although they might need advance warning.

2
Milk for Cheese

Cheese is manufactured from the milk of cows, goats and ewes. It is possible to make good cheese from all these types of milk, although some make better cheese than others.

The legal aspect of obtaining or producing milk and selling the cheese made from it are complex, especially in the case of milk from cows. Cows' milk is covered by the Milk and Dairies Regulations (1977), which state that all producers must have a licence to produce and sell milk – this licence is granted by the Ministry of Agriculture. In addition, all producers must have a quota allowance, which is the quantity of milk that may be produced before a penalty is incurred.

The milk of goats and ewes is not affected by the Milk and Dairies Regulations, and there are no restrictions on the quantity that may be produced or manufactured.

All types of milk and milk products sold are subject to food hygiene standards and will come under the jurisdiction of the local Food and Drugs Department. It is advisable to contact both the Ministry of Agriculture and the Food and Drugs Department if it is intended to sell the produce manufactured, as they will be able to give accurate information and help. There are no restrictions on products made for home consumption.

Milk Supply

Good quality cheese can only be manufactured from good quality milk. The milk must be of good microbiological and chemical quality, of normal composition and free from taints and inhibitory substances. To ensure that the milk is of good microbiological quality, the animals must be in good health with a healthy udder that is free from mastitis.

The milking equipment and premises must be of suitable design, and all milking equipment and milk receptacles must be cleaned and sterilised before use. Milk must be cooled as soon as it is produced, and then it must be stored at between 41–44.6°F (5–7°C) for no longer than two days – after two days psychrophilic bacteria (bacteria which grow at low temperature) will thrive. These bacteria produce a bitter flavour and taints in the cheese.

Small producers may freeze milk for several days until they have collected a sufficient amount to manufacture. Although this is an excellent way to preserve milk, freezing has the effect of rupturing the fat globules in the milk and this results in a greasy layer on the milk surface when the milk is thawed. This 'free' fat is lost with the whey during manufacture. To reduce this inevitable damage, it is advisable to chill the freshly-produced milk before freezing it. Channel Island milks are particularly susceptible to this problem since the globules of fat in them are larger than the other types of milk.

The chemical quality of milk is affected by the stage of lactation, the season of the year, the type of rations being fed and the general health of the animal.

The stage of lactation has the greatest effect on cheese quality. Early lactation milk (after colostrum) is best in quality since the fat and protein levels are high and good for cheese. These levels drop gradually as the lactation progresses. The milk produced during late lactation, when the protein level is at its lowest, will tend to make weak-bodied cheese with a bitter flavour. In herds where all the animals give birth at the same time, or where a producer has only one or two animals, these factors have a significant effect on the cheese.

The season of the year will dictate the type of food being fed to the animals and the best milk is produced from cows grazing on spring and early summer grass – at this time the fat and protein levels in the milk are high and the milk has a good colour. This milk is so good for cheese manufacture that many farmhouse cheesemakers run a spring calving herd in order to make best use of this grass and so produce a quality cheese. The production of good quality winter milk needs excellent conserved grass with a well-balanced supplement of concentrates. Winter milk for cheese is often inferior in quality and tends to make a weak-bodied cheese with a poor colour. An excess of certain feedstuffs, such as roots, sugar beet pulp, or brewers grains, affects milk quality and may cause taints. A balanced diet must be fed to the animals to ensure good milk for cheese.

The health of an animal has a significant effect on milk quantity. A healthy animal which is well fed will produce normal milk while a sick animal will produce abnormal milk, depending on the cause of illness. Abnormal milks do not make good cheese.

Colostrum is an example of abnormal milk and is produced in the first week of lactation. It is unstable and is not at all suitable for cheese. Late lactation milk is also unsuitable for cheese unless bulked with normal milks.

A serious cause of abnormal milk is udder disease which causes a reduction in the protein level, resulting in a soft, weak curd. The main cause of udder disease is mastitis and this is normally treated with antibiotics to clear the infection. It must be remembered that antibiotics will also destroy the cheese starter culture, so it is advisable to have a good relationship with the herdsman and to ensure that he is aware of the consequences of supplying milk containing antibiotics. If mastitis milk is used without being heat treated, there is a real risk of the bacteria causing septic sore throats in humans. Although the bacteria would not survive for very long in the cheese, the toxins produced by the bacterial cells remain viable and may cause food poisoning.

Milk for cheesemaking must be completely free from taints, as many of these will be detectable in the finished product. In the main, taints are from feedstuffs, from weeds or taken up from paint or petrol. Milk of poor hygienic quality will also be subject to bacterial taints.

Inhibitory substances in milk have a detrimental effect on the cheesemaking process. During the manufacture of cheese a starter culture of bacteria is used to produce lactic acid. If milk contains inhibitory substances, such as antibiotics, the starter culture bacteria will be inhibited or destroyed, thus ruining the cheese.

Milk Composition

The important constituents in milk for cheese are the fat, casein, soluble calcium and lactose. The cheese yield and firmness of curd are dependent on the relationship between the fat and casein while the formation of curd depends on the presence of soluble calcium. This calcium is attached to the casein (the milk protein). Lactose is the milk sugar which is reduced to lactic acid by the starter culture bacteria throughout the cheese make. The composition of the milks normally used in cheesemaking are shown in Fig 9 and Fig 10.

Cows' Milk

Most breeds of cow produce milk suitable for cheese manufacture. The best breed is said to be the Ayrshire, which produces good quality milk with a small fat globule size, ideal for cheesemaking. Channel Island breeds give milk with too high a percentage of fat for it to be ideal for cheese – the globular structure of the fat is also very large and likely to become damaged during manufacture. To help

Composition of Milk from Various Species of Mammals (percentages)

Animal	Fat	Protein	Albumin	Ash
Cow	3.75	3.40	0.4	0.75
Goat	6.00	4.00	0.7	0.84
Ewe	9.00	5.70	1.1	1.00
Buffalo	6.00	4.50	0.7	0.75
Ass	1.40	1.95	1.2	0.50
Camel	3.00	5.20	1.7	1.50
Reindeer	17.10	9.60	1.2	0.30

Fig 9

Average Composition of Cows' Milk (percentages)

Breed	Fat	Protein	Lactose	Ash
Jersey	5.14	3.80	5.00	0.75
Guernsey	4.90	3.85	4.95	0.75
Shorthorn	3.65	3.30	4.80	0.69
Ayrshire	3.85	3.35	4.95	0.69
Friesian	3.40	3.15	4.60	0.73

Fig 10

correct this it is necessary to skim the milk to reduce the fat content to about 3.8 per cent. All the cheesemaking operations must then be carried out very carefully and the milk and curd treated gently to prevent damage to the fat globules. It is possible, with some effort, to make excellent cheese from Channel Island milk and it will have the added bonus of a creamy colour and texture.

Goats' Milk

This is also excellent for cheesemaking, but it is advisable to avoid very late lactation milk as it tends to give a bitter taint to the cheese. The milk has a high percentage of total solids and produces a high yield of cheese. The colour of goats' milk cheese is very white and this may not please some customers. It is possible to improve the colour with annatto, a natural colouring agent used in cheese manufacture.

Ewes' Milk

This has a higher percentage of protein than cows' and goats' milk and, although it is a little more difficult to deal with once the curd is formed, it makes excellent cheese with a very high yield. Like goats' milk it has a very white colour, which can be improved with annatto if desired. The lactation of a sheep is relatively short compared with that of a cow or a goat, so the changes in milk composition which occur during lactation are more marked. This means that the method of making cheese from ewes' milk must be adjusted at regular intervals to allow for the changes in composition.

Expected Yields

1. Cows' milk – 1kg of cheese from 10 litres of milk (approximately, 2lb from 16 pints).
2. Goats' milk – 1.5kg of cheese from 10 litres of milk (approximately, 3lb from 16 pints).
3. Ewes' milk – 2kg of cheese from 10 litres of milk (approximately, 4lb from 16 pints).

These figures are all approximate and will vary with the composition of milk.

Heat Treatment of Milk

Ideally all milk should be pasteurised before it is manufactured into cheese. The main reasons for this heat treatment are:

1. It destroys pathogenic bacteria (bacteria that cause disease).
2. It destroys taint and fault-producing bacteria, enabling the starter culture to work uninhibited.

After this heat treatment the milk will not contain any bacteria to cause disease or infection. It is then the cheesemaker's responsibility to make sure that such bacteria do not gain re-entry to the milk during the cheesemaking process – this is done by employing hygienic methods and personal habits. The pasteurisation temperature also destroys most other bacteria, many of which would cause 'off' flavours and faults in the cheese. With the destruction of the majority of bacteria found in the raw milk, the starter bacteria have a clean environment with no competition, and can work efficiently and unhindered at producing lactic acid from the lactose in the milk.

The disadvantage of pasteurisation is that, in destroying most of the bacteria which occur in the raw milk, it destroys a few species of bacteria which are beneficial to cheese flavour. Despite this fact, it is advisable for any cheese manufacturer who is selling cheese to the general public to subject his milk to heat treatment. The benefits of pasteurisation greatly outweigh the disadvantages.

Temperatures

There are two systems of pasteurisation:

1. The milk must be retained at a temperature of 65.6°C (150°F) for at least 30 minutes and then be cooled immediately to a temperature of 10°C (50°F).
2. The milk must be retained at a temperature of not less than 71.7°C (160°F) for at least 15 seconds and then immediately be cooled to a temperature of not more than 10°C (50°F).

Note If the milk is to be manufactured into cheese immediately, then it would be cooled to 27°C (80.6°F), ready to be inoculated with the starter. It is important that the milk for cheese should not exceed the recommended pasteurisation temperature, because a higher temperature reduces the amount of soluble calcium in the milk, and this will affect the firmness of the curd.

Methods

Small-scale production could make use of small pasteurising plants such as a plate heat exchanger, or a holder tank. If the quantity of milk will not allow for such equipment the milk can be heated in a double pan and cooled in a sink of iced water. Some cheese vats have a heating and cooling facility incorporated.

Post-pasteurisation Contamination

Care should be taken to ensure that all cheesemaking procedures are carried out hygienically to ensure that heat treated milk is not contaminated during the manufacture of the cheese.

3
Starter

A good starter is the key to good quality cheese and this chapter is, therefore, particularly important.

Starter is a culture of lactic organisms which is added to the warm milk at the beginning of the cheesemaking process. These organisms reduce the lactose in the milk to lactic acid, to give a 'clean wholesome souring'.

The production of acid is essential to the cheesemaking process. The acid is partially responsible for the flavour of the cheese, and for the changes in the texture of the curd which occur during the manufacture, storage and ripening of the cheese.

Requirements of a Good Starter

1. The culture must produce acid at a controllable and consistent rate.
2. The culture must produce a clean acid flavour.
3. The culture must be free from contaminants.
4. The culture must tolerate the scalding temperatures required, according to the type of cheese being made.
5. The culture must tolerate the level of salt in the stored cheese.

Starter Strains

The species of bacteria which are used in the manufacture of most hard cheeses are strains of *Streptococcus lactis* and *Streptococcus cremoris*. These bacteria are able to reduce one per cent of lactose to one per cent of lactic acid, with no other substances being produced – in this way, a clean acid flavour is given to the cheese. Occasionally *Streptococcus diacetylactis* is used, and also some of the *Leuconostoc* species of bacteria. These particular organisms produce acid and flavour substances in cheese which are not always considered desirable. The degree of flavour substances produced by these species varies between each batch of cheese, and this makes the grading difficult when large quantities are involved. Small producers can make good use of starters that include these species because they produce a cheese with an interesting flavour which will appeal to the specialist market.

Starter cultures are composed in two ways. They can be made up of a single strain of lactic bacteria ('single-strain starters') or of a mixture of strains ('mixed-strain starters').

Single-strain starters are normally a culture of a single strain of *Strep. lactis* or *Strep. cremoris*. This type of starter has the advantage that acid production is predictable and steady throughout the cheese process. The disadvantage of a single-strain starter is that it is liable to suffer from a bacteriophage attack and the starter culture will fail, resulting in a loss of cheese. Bacteriophage is a bacterial virus which attacks and destroys bacterial cells.

Mixed-strain starters are usually cultures of two or more lactic species, for example, *Strep. lactis*, *Strep. cremoris* and *Strep. diacetylactis*. A mixed-strain is more unpredictable during manufacture, but has the advantage that in the event of a phage attack the starter may slow up, but will not fail completely.

Growth of Starter Cultures

Milk is an ideal medium for starter growth because it contains water, lactose and trace elements. Milk also contains dissolved oxygen, and, although starter organisms require some oxygen, they grow best in an environment where it is reduced. From a practical viewpoint, once the starter has been added to the cheese milk and stirred in well, it is not necessary to stir the milk continually during the ripening period. Excessive aeration will, in fact, reduce the rate of acid production, and in this case it may be necessary to extend the ripening time.

The Effect of Temperature

The optimum temperature for starter growth is between 27 and 30°C (80.6–86°F). Temperatures lower than 21°C (70°F) will inhibit bacterial growth during the initial ripening period of the cheese, and cause a slow cheese in the initial stages of the process. Temperatures higher than 30°C (86°F) are outside the optimum range and will not encourage efficient growth. At temperatures of 40°C (104°F) and above, the starter bacteria are destroyed, although some strains of lactic organisms are more tolerant of high temperatures than others.

Causes of a Slow Starter

A slow starter culture is one that does not produce acid at a fast enough rate – this can be due to several causes.

Inhibitory substances in the milk will cause the starter to be slow, and the presence of antibiotics may even cause the starter to fail completely. The cheesemaker would be aware of a slight lack of starter activity up until the scalding stage. After the scald the cheese is stirred for an hour, and during this time lack of acid production becomes very apparent. The cheese may progress at a slow rate or the production of acid may stop altogether. This pattern is fairly typical of starter failure due to antibiotics. The presence of antibiotics in the milk is normally an accidental occurrence, so that slow cheese will probably only be a problem on the one day.

Bacteriophage (phage) is a bacterial virus – each strain of bacteria has a specific phage which will attack the bacterial cell and destroy it. A phage is difficult to diagnose and the signs may be confused with the effect of antibiotics. The production of acid during the cheese make will seem normal, or appear to slow up only very slightly, until the whey is drained off. After this, the rate of acid production slows down rapidly and eventually stops. It may be possible to rescue the cheese, but it will then be slow to mature and extremely weak in body. It is advisable to check that the starter propagation technique is being carried out hygienically.

A practical approach to phage prevention is to make sure that all milk and whey residues are completely removed from the cheeseroom, and that all drains are thoroughly cleaned. Phage is easily destroyed by chlorine, so the use of a dairy steriliser such as Sodium Hypochlorite, after the equipment and floors have been thoroughly cleaned, is one of the best ways to prevent infection.

Another cause of slow acid production is the use of old starters which may have been in the deep freeze for longer than six weeks, or have been subcultured for over six months. It is advisable to renew the mother cultures ('stock cultures') regularly to ensure that starter activity is maintained.

Poor starter routine is usually a prime cause of slow starter or starter failure. The starters become contaminated and this results in 'off' flavours, poor curd texture and production of gas in the finished cheese. The production of 'off' flavours and gas are more noticeable in soft cheese.

Milk of poor microbiological quality will also have a slowing effect on starter activity, because the contaminating bacteria will inhibit the growth of the starter bacteria and cause faults and 'off' flavours.

Starter Production

Milk for Starter Production

The milk for starter production should come from the same animals as the milk used for the cheese, and its source should be chosen with great care to ensure that the growth of the cultures will not be impaired. It is essential that the milk should be completely free from inhibitory substances, and the best way to make certain of this is to take the milk from a selected animal whose recent history of veterinary treatment is well known. Many cheesemakers have a particular animal whose milk they favour for the use of starters.

An alternative source of starter milk is dried skimmed milk which is usually reliably free of antibiotics. The starter cultures will grow very well in a solution of milk made up so that the total percentage of solids is between 9 and 11 per cent. With goats' and sheeps' milk cheese, it may be advisable to use dried milk of the appropriate type to ensure the purity of the cheese. The reliability of dried goats' and sheeps' milk is questionable – although the country of origin will be clearly indicated on the package, there is very often no indication as to its freedom from antibiotics, or from other substances which may be undesirable for starters. Care must be taken when using these dried milks and several trial runs with starter cultures are recommended before taking them into general use.

Starter Cultures

The cultures used in small-scale production are:

1. The mother culture – regularly subcultured and grown in freshly sterilised milk.
2. A working culture – a larger quantity of milk which is used to inoculate the cheese milk.

Preparation of the Mother Culture

Equipment:
Pressure cooker
Heat resistant glass bottles with screw caps
Skim milk
Pure culture

Fig 11   Starter milk can be prepared by heating milk in a domestic pressure cooker. The milk is contained in a glass bottle, with a loose fitting cover to allow for the expansion of liquid during heating.

Preparation of Milk   All milks used for starter propagation must be sterile, and this may be achieved by the following method. Pour the milk into clean glass bottles to about two-thirds full, and lightly screw the caps on to the bottles – do not tighten them. Cover the base of the pressure cooker with water and place the bottles of milk inside. Pressure cook the bottles of skim milk at 15psi (high pressure) for ten minutes and then cool the bottles gradually before tightening the caps. Starter milks can be sterilised in batches and stored until needed in the refrigerator or a cool cupboard. Over-sterilising the milk until it is brown will have caused the lactose to caramelise. This cannot then be used by the starter bacteria and the activity of the culture will be impaired. If the milk does tend to become caramelised reduce the pressure to 10psi (medium pressure) and increase the 'cooking' time to fifteen minutes.

Inoculation of the Milk   Using as clean a technique as is possible, transfer three or four drops of the pure culture into a fresh bottle of sterilised milk, taking care to close the bottles immediately after the

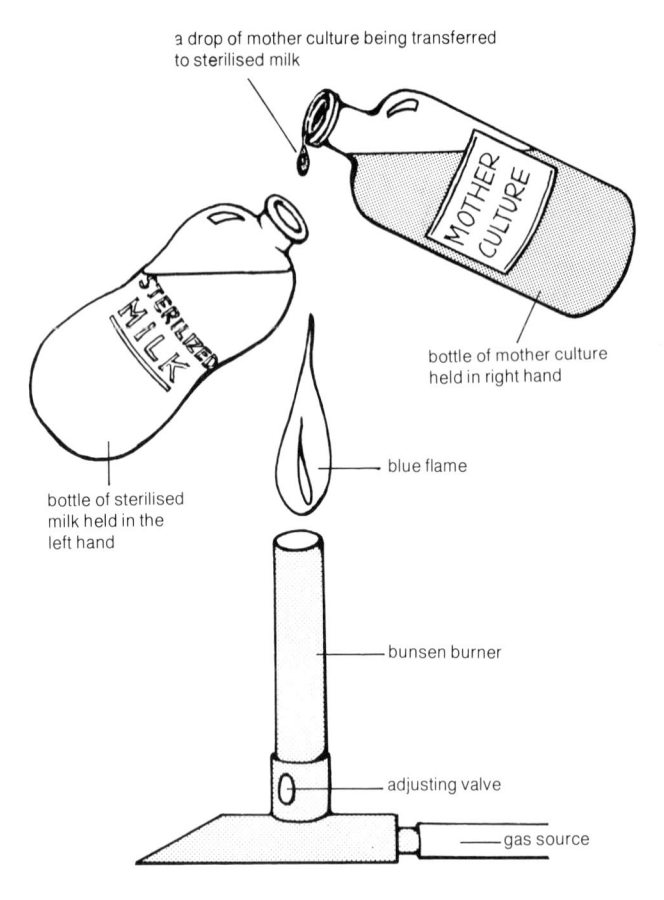

Fig 12 Starter propagation is carried out aseptically by transferring a drop of mother culture into freshly sterilised milk over a flame.

transfer. If it is possible to carry out the inoculation over a Bunsen flame there is less risk of contamination. The inoculated milk is then incubated for twelve hours at 24°C (75°F). An airing cupboard, or similarly warm place, will be suitable, provided that the temperature is consistent.

When cheese is made daily or several times a week, the mother culture must be subcultured every day. If cheese manufacture is less regular, the following procedure can be followed.

Prepare and sterilise several bottles of skim milk for inoculation and inoculate, as aseptically as possible, all the fresh bottles with the pure culture. The bottles of inoculated milk are then frozen instead of being incubated. When the mother cultures are required a bottle can be defrosted and, after incubation for six to eight hours at 24°C (75°F), it will be ready for use. If this method is used do not store the inoculated bottles for more than a month – the activity of the culture will be reduced and the quality of the starter poor.

It is important to note that when sterilising milks for freezing, the mother culture bottle must not be more than two-thirds full otherwise the glass will crack.

Preparation of the Bulk Starter

This procedure should be carried out the day before the bulk starter is required.

Equipment:
A large saucepan or preserving pan
A milk can with a well-fitting lid
A piece of flat board which fits into the base of the pan
Whole or skim milk

Preparation of the Milk Pour the fresh milk into a suitable can and replace the lid. Place the can on top of the wooden board in the saucepan. The saucepan is then filled to about two-thirds full with clean water which is brought to the boil. The water is then adjusted to a simmer, and the milk kept at this temperature for 45 minutes before being cooled to 27°C (80.6°F) ready for inoculation. If the milk is not going to be inoculated immediately, cool the milk further and keep refrigerated until needed. It will then be necessary to re-heat the milk to 27°C (80.6°F).

The starter milk must be carefully inoculated with about two per cent of a newly-incubated mother culture and the can recapped.

The timing of the inoculation and incubation should be organised so that the cheesemaking starts at the end of the twelve-hour incubation period. This is important because the starter bacteria are at a stage of peak activity after twelve hours – it is at this stage of growth, therefore, that they should be inoculated into the cheese milk.

The acidity of the starter will be within the range of 0.75 per cent–0.90 per cent lactic acid after twelve hours incubation.

It is important to sniff the incubated starter as the can lid is being removed. The starter should give off a sharp, clean, acid smell which

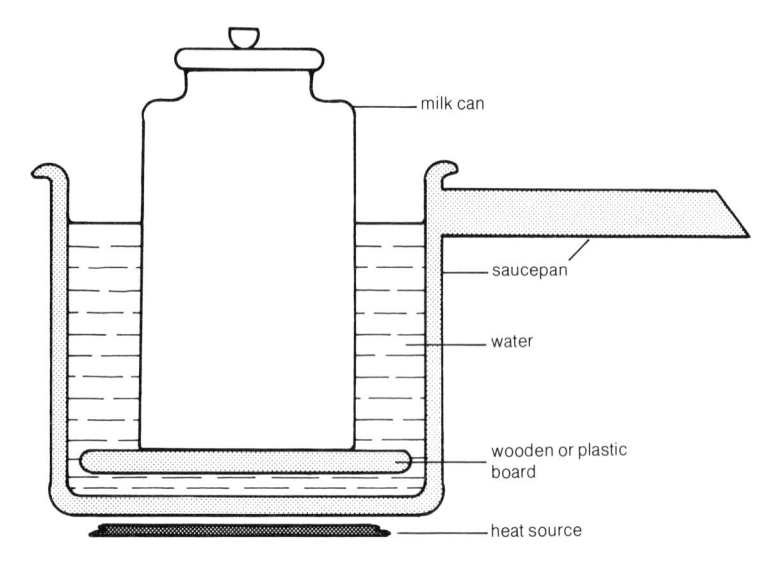

Fig 13 Bulk starter milk is prepared by placing the can of milk on a board in a saucepan of water. This is a cross-section of the positioning of the can.

is very distinctive. With experience the cheesemaker will be able to detect 'off' flavours and know when the starter is contaminated with other bacteria, or is not as active as it should be.

Any doubt about the viability of the starter will require a decision as to whether to use it or not. Generally, it is not advisable to use a doubtful starter and it may be necessary to resort to using DVI (Direct Vat Inoculation) culture for that particular day's make. This is a culture which is added directly into the vat at the start of the cheesemaking process.

Direct Vat Inoculation Starters

DVI starters have been available to commercial cheesemakers for several years. They are cultures of starter bacteria which have been freeze-dried and packaged in such a way that they can be stored in a domestic refrigerator for a limited period of time. These starters are now available for small quantities of milk, and are very useful as a reserve in case the liquid starter fails. They are expensive, but are very reliable and require no preparation.

New cheesemakers are not recommended to use DVI until they

have gained a great deal of experience, because the making of the cheese has to be adjusted to accommodate the different type of culture form.

Sources of Cheese Starters

Starter cultures should be purchased from reliable sources to prevent the risk of starter failure or contamination. Liquid starters are available from teaching establishments and from commercial suppliers, most of whom will despatch by return of post. Freeze-dried mother cultures can be obtained from several manufacturers and DVI cultures are available from a French company, Eurozyme (*see* Useful Addresses).

Storage and Records of Starters

All starter cultures and prepared milk, unless in use or in a freezing cabinet, must be kept refrigerated. Each starter should have a number or name by which it can be identified and a record should be kept. The record can be contained in a small notebook and should include the identification of the starter, the date the pure culture was purchased and the details of starter behaviour during the cheesemaking process. The record should also be tied in with the quality of the cheese when ready for consumption.

4
Rennet and Additives

Rennet

The action of rennet in warm milk is one of the most fascinating stages of the cheesemaking process. The rennet contains an enzyme which is responsible for the coagulation of the curd during the early stages and it is also responsible for some of the changes in the curd texture and flavour during manufacture and storage of the cheese. Rennet contains the enzyme *rennin* – its action in milk is very complex, so the following explanation is simplified.

Rennet Action

Milk contains fat, proteins, lactose, water and trace elements. The protein which is important to the cheese is the casein, and attached to the casein is a soluble salt called calcium phosphate. These two constituents together form calcium caseinate which exists in milk as a complex molecule. The casein molecules are evenly dispersed throughout the milk when it is fresh, because they have reactive 'arms', and these 'arms' cause them to repel each other.

When the rennet is added to the milk the enzyme rennin is absorbed on to the 'arms' of the casein molecules. This causes the molecules to change, and, instead of repelling each other, they are mutually attracted. Their 'arms' link up, gradually forming a gel. As the gel forms (shown by the milk coagulating), all the other milk constituents are trapped within it and curd is produced. The formation of curd is shown in a simple form in Fig 14.

Rennet Use

It is important to have an insight into the action of rennet because timing at this stage of the process is critical. From the time the rennet is added to the milk until the curd begins to form there is a lapse of four minutes. The cheesemaker must be sure of several factors:

(a) casein dispersed in raw milk

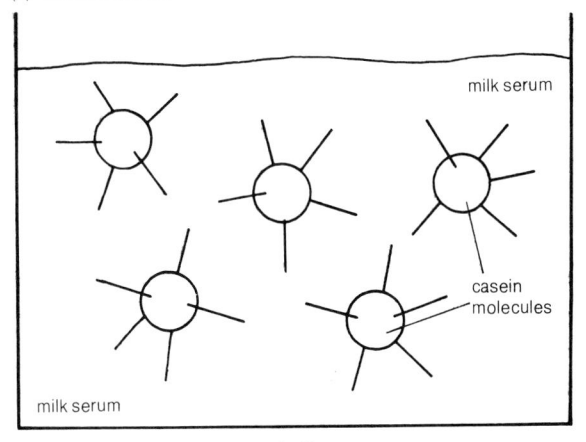

milk serum

casein molecules

milk serum

vat of milk

(b) denatured casein

casein molecules

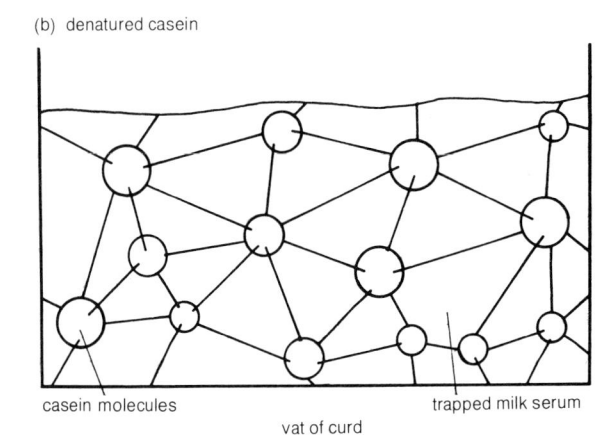

trapped milk serum

vat of curd

Fig 14 (a) Casein molecules exist as single units dispersed throughout the milk medium. (b) After rennet addition these molecules bond together forming a three-dimensional structure. This three-dimensional structure traps the milk constituents and curd is formed.

Fig 15 Rennet diluted with cold water is added to the warm milk. The rennet is distributed evenly throughout the vat or bowl.

1. The milk temperature must be 30°C (86°F).
2. The milk should be slightly acidic – 0.18 per cent LA.
3. The rennet solution must be evenly distributed and well stirred in during the four minutes.
4. Stirring must stop at four minutes.

The milk temperature is set at 30°C (86°F) because the curd will set firmly at this temperature in about 45–60 minutes (perhaps a little longer in cold weather). Temperatures higher than this will interfere with the scalding process of the cheese and also with starter growth. Temperatures lower than this will result in a curd which takes too long to set and is too soft to cut.

The more acidic the milk the quicker the curd will set, so when the cheesemaker uses a liquid culture as starter, the acidity at renneting will be about 0.18–0.19 per cent lactic acid. If a DVI culture is used the acidity will be the same as the initial milk, that is, 0.14–0.16 per cent lactic acid. The milk will form a softer curd at this acidity.

The rennet is normally used at a standard rate of 30ml in 100 litres of milk (1fl/oz in 20 gallons). To ensure even distribution and rapid mixing when it is added to the milk, it is advisable to dilute the rennet

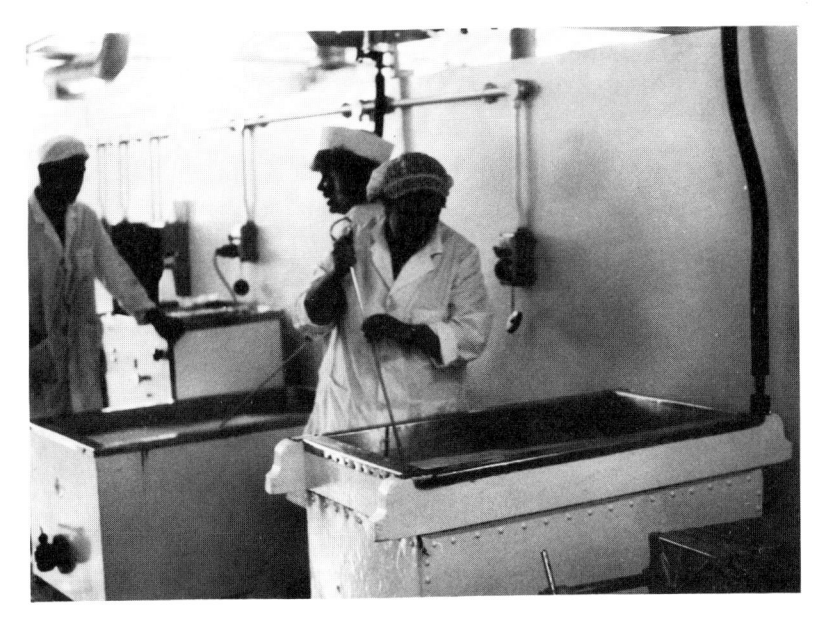

Fig 16 The rennet is stirred into the milk immediately after addition and stirring is continued for four minutes, whatever the volume of milk. Stirring must be carried out smoothly and evenly and stopped after the appropriate time.

in about six times its volume of *cold* water. Hot water will destroy the enzymes.

Rennet in Various Types of Milk

Rennet action in the milk used for cheesemaking is basically the same for all types. However, the volume of rennet used has to be adjusted for the milk of the different animals because the composition of each is not the same.

The composition of the milk of most breeds of cow is such that rennet used at the standard rate is correct. For breeds which have a higher percentage of total solids in their milk it may be necessary to reduce the amount of rennet slightly. These breeds include the Channel Island cattle and the South Devon. 25ml in 100 litres of milk (less than 1fl/oz in 20 gallons) may be sufficient, but it will perhaps be necessary to increase the volume if the curd is soft.

The total solids in goats' milk is higher than in cows' milk, so a

RENNET AND ADDITIVES

reduced rennet usage is recommended. The quantity could be as low as 20ml in 100 litres (approximately ⅔fl/oz in 20 gallons) for goats' milk in early to mid lactation, but towards the end of lactation it may be necessary to increase it to 25ml in 100 litres (just under 1fl/oz in 20 gallons).

Ewes' milk has the highest total solids of all the milks used, and the rate of rennet used can be as low as 10ml in 100 litres (⅓fl/oz in 20 gallons) during early lactation, gradually increasing in volume to 15ml in 100 litres (½fl/oz in 20 gallons) later in lactation.

Types of Coagulant

Calf rennet is regarded as the most important rennet for cheesemaking, containing about 90 per cent pure rennin. It is obtained from the fourth stomach of milk-fed calves and is then purified and sold, normally in liquid form. Rennet is also available as a powder and a tablet, and these are useful in warmer climates where liquid rennet does not keep very well.

The original farmhouse cheesemakers used a calf vell – the dried fourth stomach of a calf (abomasum) – as a source of rennet. The vell was immersed in cold salt water for a few minutes, to allow some of the enzymes to dissolve and the resulting solution was used to coagulate the milk. The vell was then pegged out in the air to dry.

The other forms of rennet are mixtures of pure rennin and various forms of pepsin, and these are known as '50/50 rennets'. Their disadvantage is that they produce bitter flavours in cheeses that are kept for longer than six months.

Microbial rennets are available and are suitable for the manufacture of vegetarian or kosher products. The microbial rennets are extracts of enzymes similar to rennin, which are produced by certain moulds and bacteria. An example of one of these moulds is *Mucor miehei*. The microbial rennets are normally purchased as a liquid and are used in a similar way to the standard rennets – like the 50/50 rennets, microbial rennets tend to produce bitter taints in long-keeping cheese.

Plant rennets are not easily available, and very little research has been carried out into them. It has been found that most of the plants which have been used traditionally do not actually coagulate the milk, but cause the curd to separate out, as in souring. The plant rennets which do exist tend to come from the Asian countries, and are derived from the fig tree (*Fiscus carsica*) and the paw-paw (*Carica papaya*). The addresses of some suppliers of rennet will be found at the back of this book.

Storage of Rennet

All rennets must be stored in cool conditions. Liquid rennets should be refrigerated and other forms stored as directed. It is advisable to be familiar with the smell of fresh rennet and to check all rennet before use. Do not use rennet which looks cloudy or does not have a typical smell.

Annatto

This is the only permitted food colouring for cheese according to The Cheese Regulations (1970). It is a carotenoid pigment which is derived from the shrub *Bixa orellana* which grows in the West Indies. The colouring is a natural substance, supplied in a liquid form and it can be used to colour cheese or to improve the colour of winter milk used for cheese. The English cheeses which are coloured are Leicester, Double Gloucester and Cheshire. Annatto has an interesting action in the milk because the colour intensifies as the acidity of the cheese increases. There is no cause for worry if the milk looks rather pale when the annatto is added at the beginning of the process, as the colour will improve throughout the cheese make and during storage.

Use of Annatto

The annatto must be well stirred into the milk at least 15 minutes before the addition of the rennet to ensure that it is well mixed in before the curd forms. If mixing is poor the finished cheese will have a streaky appearance.

Rates of Use

1. For a medium-coloured cheese 60ml in 450 litres (2fl/oz in 100 gallons).
2. For a deep coloured cheese 100ml in 450 litres (3fl/oz in 100 gallons).
3. For improving the colour of winter milk 15ml in 450 litres ($\frac{1}{2}$fl/oz in 100 gallons).

The strength of colour in each batch may vary a little and the age of the annatto will also affect the colour strength – it may be necessary, therefore, to adjust these quantities at times to achieve the desired depth of colour in the cheese. When using the annatto to improve milk colour it will be necessary to adjust the quantity used according to the quality of the milk, and in particular the rations being fed to the stock. The natural colour of the milk is usually poor (due to the lack of carotene) whilst the animals are fed conserved fodder during the winter months; it improves rapidly when the animals go out to grass in the spring.

Salt

Salt is an essential ingredient in almost every variety of cheese and has several functions:

1. It improves the flavour of the finished cheese.
2. It helps control the growth of the lactic bacteria towards the end of the cheesemaking process.
3. It aids the drainage of the curd after milling by causing the curd to shrink.
4. It improves the keeping quality of the cheese by suppressing the growth of undesirable organisms.

The method of adding salt depends on the variety of cheese being made. Cheddar, Cheshire, Double Gloucester and most other hard cheeses are dry-salted at the milling stage of the process. The salt is distributed evenly over the milled curd and mixed in well.

Caerphilly is an example of cheese that is brined after pressing. The cheese is immersed in a 20 per cent brine solution. This method of salting forms a rind on the cheese and eliminates the need to bandage the cheese surface. It can only be used on cheese with a short maturing period.

Soft cheeses are very often salted by rubbing salt gently into the surface. The salt gradually permeates the cheese, over a period of about 24 hours, and it is then ready for consumption.

Rates of Use

The salting rate will depend on the variety of cheese being made. Generally a rate of two per cent by weight of cheese is used. It is sufficiently accurate to calculate the weight of salt needed according to the estimated yield of cheese. The yield of cheese will depend on the type of milk being used and its volume.

Quality and Storage

The best type of salt is pure vacuum-dried sodium chloride. This will mix into the cheese evenly and will not cause faults due to impurities. For small quantities of cheese, household salt is adequate. However, it contains magnesium carbonate to prevent it caking and is not always considered desirable.

The storage of salt is important, and care should be taken to ensure that it remains clean and dry. A suitably sealed plastic container or plastic dustbin will make a good store.

Additives

Cheesemakers very often like to give their products extra value and this can be achieved by using additives such as herbs, spices, nuts and alcohol. Smoking is another method of giving cheese a different flavour.

The herbs must be clean, and can be used fresh or dried. Herbs often used in cheese are sage, mint, lavender, tarragon, fennel and chives. Spices, including caraway seeds, cloves, nutmeg and peppers, can be used – these must be purchased from a reliable source to prevent contamination of the cheese due to undesirable organisms. Herbs and spices are normally mixed into the cheese after the fresh curd is salted, and before it is moulded and pressed.

Nuts are always added to mature cheese. The cheese is remilled, and the chopped nuts are mixed into the curd before it is remoulded and pressed. It is important to blanch the nuts with boiling water before adding them to the curd. Nuts added to fresh curd tend to decompose during the cheese maturing process, resulting in a very unpleasant product.

Alcohol has been added to cheese for centuries, most notably in the dosing of Stilton cheese with port. Alcohol can be added to the fresh curds or it can be added to the finished cheese. Some cheeses are dosed by immersion in alcohols such as beer and wines.

Smoking has become a popular method of flavouring cheese in recent years. This is done by hanging the cheese in a smoke room. The smoke is produced from oak or applewood shavings and has the effect of evaporating moisture and bringing fat to the surface of the cheese. The fat combines with substances in the smoke and this has a preservative effect on the cheese, provided that it is kept dry.

5
Hygiene, Water Supply and Effluent

Hygiene

This chapter is very important to the cheesemaker. The cheesemaking process is based on the production of acid from a pure culture of lactic bacteria. If the standard of dairy hygiene is poor, many other species of bacteria will gain entry to the cheese. These adventitious bacteria may produce acids, gases and many other substances, resulting in undesirable flavours and serious faults in the cheese. It is, therefore, important to maintain a high standard of hygiene in the dairy.

The Dairy

The ideal dairy building should be light, airy and warm. The construction should be such that all surfaces are easily cleanable. The floors must have an even gentle slope to a trapped drain and the floor must be impermeable to water and have a non-slip surface. The walls and ceilings must have a non-porous surface that is white or cream in colour – a light colour is recommended because it makes the build-up of dirt more noticeable. Windows and doors must seal properly and have surfaces that are easily cleaned. It may be advisable to have a movable fly screen which can be inserted into the window space during the summer months when the windows are open.

Furniture in the dairy should comprise a large stainless steel sink for washing equipment, and a small hand basin. These must have a plentiful supply of hot and cold water. A hose pipe facility is very useful, but there must be a storage rack or hook to keep it from lying on the floor, where it would be a hazard to the cheesemaker.

Cleaning the Dairy Daily cleaning involves scrubbing the floors and drains, and wiping or scrubbing all areas of wall susceptible to splashes of water or milk product. Weekly cleaning involves scrubbing all the walls, windows and doors. The equipment must be cleaned daily to a high standard.

Dairy Personnel All dairy workers should wear a clean overall and hat and the hat should cover all the hair. Rubber boots must be worn and kept in a clean state. Shoes are not recommended for wear in the dairy as the cheese whey quickly rots them.

Personal hygiene must be of a high standard. Arms and hands must be thoroughly washed before touching the product, and particular care must be taken after a visit to the WC.

If a dairy worker is suffering from an infectious condition, such as a sore throat or diarrhoea, they must not work with dairy products until they have recovered.

To comply with these requirements it will be necessary to have a supply of clean overalls and hats, since these become wet and sticky during cheesemaking. A non-perfumed soap must be provided, and a supply of clean towels – paper towels are very hygienic as long as there is a foot-operated disposal bin available. The WC must be in a separate building with separate washing facilities.

The Kitchen

This is just as important as hygiene in a purpose-built dairy, and it is usually more difficult to achieve. The kitchen must be scrupulously clean, and, if possible, an area should be set aside for making cheese. All equipment and utensils used in the cheesemaking should be kept apart and employed only for this purpose.

Under no circumstances must raw meat be handled in the kitchen whilst cheese is being made. Raw meats (particularly poultry), are a source of the bacteria that may cause food poisoning, the most serious of which are the *salmonella* species. Cheese is not cooked at a temperature high enough to destroy the *salmonella* or any other pathogenic bacteria, and these would survive in the finished cheese for some considerable time.

All equipment used for the cheese must be washed up separately from the normal kitchen utensils.

Equipment and Materials for Cleaning

Equipment needed for cleaning is as follows:

A long-handled broom for the floor.
A long-handled squeeze blade for the floor.

Turk's head brush for the drain.
Turk's head brushes for general washing up.
Small brushes for pipes and outlets.
Green nylon scouring pads.
Buckets (metal and plastic).

It is important to purchase good quality equipment, paying special attention to the brushes. Poor quality brushes lose their bristles and it would be rather unfortunate if a customer were to find one of these in the cheese. The list does not include dish cloths. These are a serious source of contamination and should be used with care. Always use a fresh cloth and ensure that all cloths are washed and boiled at the end of the day.

The materials needed for cleaning are very simple – a detergent and sterilising agent. The detergent must be specifically designed for use in the dairy – it must clean surfaces, rinse off easily and it must be odourless. This type of detergent is available from most agricultural suppliers.

The sterilising agent must be Sodium Hypochlorite. This agent is used because it contains chlorine which will destroy all the bacteria on a surface, provided that the surface is thoroughly clean. Hypochlorite is also important for use in dairies because it is effective in destroying the phage virus which attacks the starter culture. It, too, is available from agricultural suppliers.

Cleaning Methods

Do not let washing-up accumulate throughout the day. Utensils which are still milky or covered in whey are ideal breeding grounds for the phage virus.

The sequence of operations used for cleaning dairy utensils and equipment is the same for everything:

1. Rinse off all the residual product with cold water.
2. Scrub with warm or hot detergent.
3. Rinse off the detergent with clean water.
4. Sterilise the equipment with a hypochlorite solution.
5. Rinse with clean water.

The quantity of detergent necessary will be shown in the manufacturer's directions on the side of the container. The quantities of hypochlorite vary according to requirements:

To sterilise equipment prior to use requires a mixture of 30ml in 45 litres of water (1fl/oz in 10 gallons of water).

Sterilising equipment after use requires a mixture of 120ml in 45 litres of water (4fl/oz in 10 gallons of water).

Preparation of Equipment

There are two methods of preparation. The first is to scald all equipment with boiling water. This is done by placing the utensil needed in a bucket and, using another bucket or measuring jug, pouring on the boiling water. When the utensil has cooled sufficiently to handle, it is ready for use. This system of scalding will destroy the coliform bacteria which are responsible for causing gassiness and a 'cowey' flavour in cheese of all types.

The second method is to use a solution of hypochlorite, rinsing the equipment in the solution prior to use. A solution equivalent to 30ml of hypochlorite in 45 litres of water (1fl/oz in 10 gallons) is sufficiently effective against the type of bacteria it aims to destroy.

Fig 17   Equipment can be prepared for use by scalding with boiling water using a kettle or other source.

Water Supply

A cheese dairy needs a plentiful supply of water, and this must be of excellent microbiological quality to ensure that the cheese is not contaminated during manufacture.

Sources of water can be the mains supply, or a well, spring, or borehole. The mains supply is safe for use with milk and milk products but wells and springs are subject to pollution and their bacterial quality should be checked at regular intervals. Care must be taken to protect wells and springs from sources of contamination, such as drainage gulleys or stray animals. Boreholes are usually a good source of clean water because the water is drawn from deep down in the earth. It may, however, be necessary to check the water occasionally to be sure of its purity.

Water sampling may be undertaken by the local Environmental Health Officer or the Water Authority.

Effluent

The effluent produced from cheesemaking is whey. It is important to consider how to dispose of whey, as it must not be allowed into the normal drainage system. It has a high biological oxygen demand (BOD) – higher than silage effluent – and causes serious pollution if allowed to run into ditches and waterways.

Whey from cheese produced in the kitchen can be fed to stock, used to water the garden, or used in the making of bread. (It is an excellent ingredient for this.)

Whey from cheese produced on a larger scale can be fed to stock – fresh whey to the cows, and fresh or stored whey to pigs and other stock. In all cases the animals develop a shiny 'bloom' to their coats, which is due to the vitamin B_{12} found in the whey. An alternative is to use it as a fertiliser and spray it on the fields. It is important not to spray near a water course – in the event of a water course being contaminated inform the local Water Authority who will give advice.

6
Control Tests and Cheese Records

The control of the cheesemaking process is carried out by determining the acidity and temperature throughout the process, and by measuring the temperature and humidity during storage.

Acidity can be measured by using the acidity test, a pH meter or the hot iron test.

The Acidity Test

The acidity measured is the 'titratable acidity' (*see* Appendix). The test is carried out by using an alkali of known strength to neutralise the acid present in a measured volume of milk or whey. The result of the test is shown by using an indicator which turns pink at the point of neutral.

This test is used by most cheesemakers and is easy to carry out with a little practice. The procedure is as follows:

Equipment   10ml pipette
10ml squeeze-fill burette
White porcelain basin
Stirring rod
A dropping bottle or 1ml pipette (squeeze-fill)

Reagents   N/9 Sodium Hydroxide
Phenolphthalein (indicator)

Method   Suck milk into the pipette, and before it drains out put the end of the forefinger firmly over the top of the pipette. Ease the finger off slightly until milk drops out of the base. Allow the milk to drop until the level of 10ml is reached. At this point dispense the remaining milk into the white porcelain dish.

Check that the burette of N/9 Sodium Hydroxide is filled, by

CONTROL TESTS AND CHEESE RECORDS

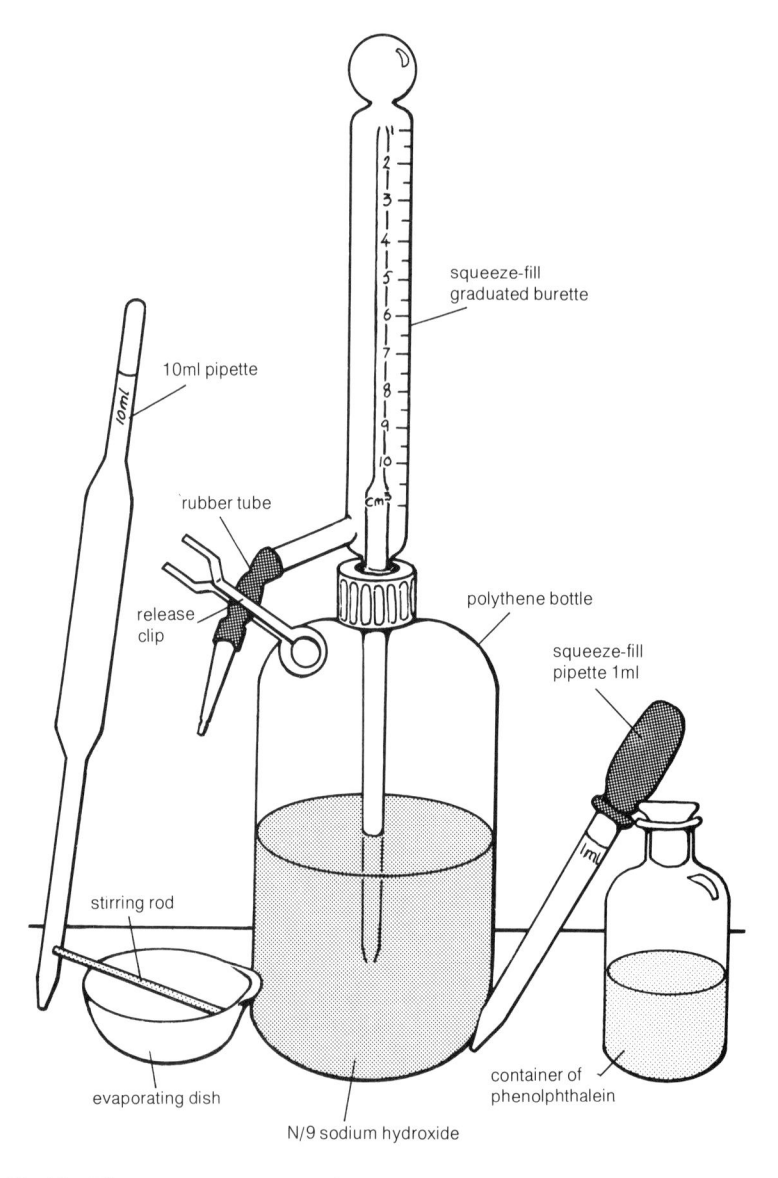

Fig 18 The equipment required for the acidity test includes a squeeze-fill burette containing N/9 Sodium Hydroxide, a 10ml pipette, a bowl and stirring rod, phenolphthalein and a squeeze-fill pipette to dispense it.

Fig 19 The equipment shown is used to carry out acidity tests during cheesemaking. The milk or whey sample is measured into the white dish with the indicator and the sodium hydroxide is run in from the graduated burette.

squeezing the plastic bottle, and place the dish under the outlet.

Add the phenolphthalein indicator to the milk using either 1ml from a squeeze-fill pipette or 2–3 drops from a dropping bottle.

Run the sodium hydroxide into the milk by squeezing the metal clip on the rubber tube, stirring continuously. Do this very carefully but not too slowly. When the colour of the milk begins to turn pink, stop running the sodium hydroxide, and read off the burette scale the volume of hydroxide used.

The reading will be in whole numbers but the actual acidity will be recorded as a decimal number. For example, a reading of 2.4 will be recorded as 0.24 – for the reasons for this *see* the Appendix.

This test is very simple in principle and with practice can be carried out quite quickly.

The pH Meter

A pH meter can be used to ascertain the acidity of the cheese, but it has certain limitations. It is a sensitive instrument which is expensive to purchase, and a hot, wet, sticky cheeseroom is not the ideal place in which to use it – there is a real risk of damage.

The use of a pH meter must be according to the manufacturer's instructions and great care should be taken both in its use and its storage in the cheeseroom. A scale comparing pH readings with acidity can be found in Fig 20.

The Hot Iron Test

This is a test which could be described as 'ancient history' in cheese terms. It was a test used in farmhouses and creameries before the acidity test became commonplace and it can only be used from the cheddaring stage onwards. Up until the cheddaring all acidities had to be estimated by smell, taste and the sensitivity of the cheesemaker.

A flat metal bar is used, about 20cm (8in) long with a wooden handle at one end. It is heated red-hot and allowed to cool to black heat. A representative sample of curd – about $4cm^3$ ($\frac{1}{4}in^3$) – is then pressed firmly onto the hot surface and allowed to cook on. When the curd is gently drawn away, fine threads of protein are seen between the iron and the curd. The length and firmness of these threads are an indication of the curd acidity – the greater the acidity the longer the threads can be stretched before they break:

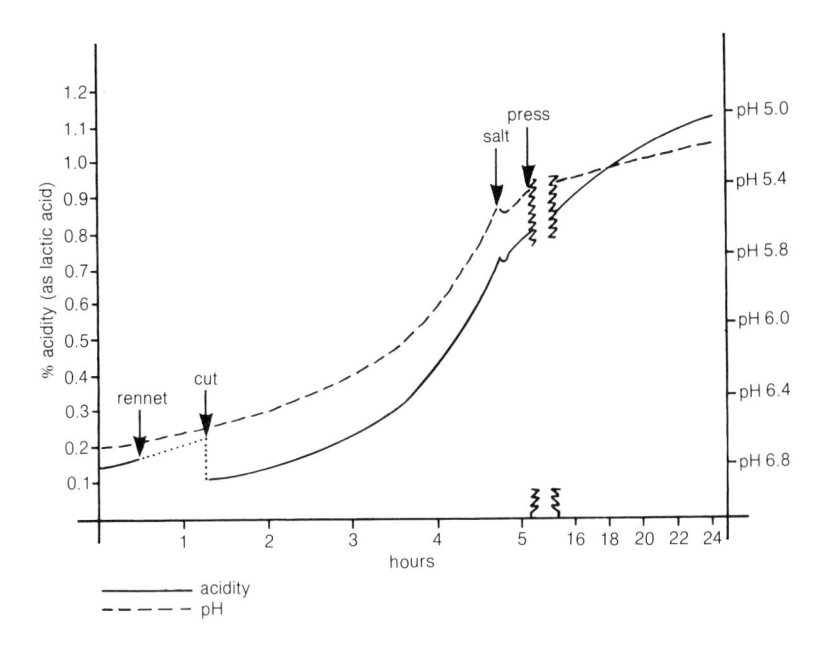

Fig 20 The graph shows the comparison between acidity and pH measurements during a cheese make. (The continual line represents the acidity readings and the dotted line the pH readings.) The difference is due to the fact that acidity readings have to be taken from liquid samples i.e. from milk or whey; pH readings can be taken from solid samples i.e. curd, and these will have a higher natural acidity.

1.25cm (½in) is equivalent to 0.25 per cent lactic acid (LA)
2.5cm (1in) is equivalent to 0.65 per cent LA
3.75cm (1½in) is equivalent to 0.75–0.85 per cent LA

This test is a useful aid to cheesemaking, but it is dependent on the skill and experience of the cheesemaker in interpreting the results.

Temperature Measurement

Temperatures are measured using a dairy thermometer which will float in milk or whey. To ensure that the thermometer is not broken whilst in use it is provident to use a plastic cover, especially designed to protect it. Thermometers are available with mercury or alcohol

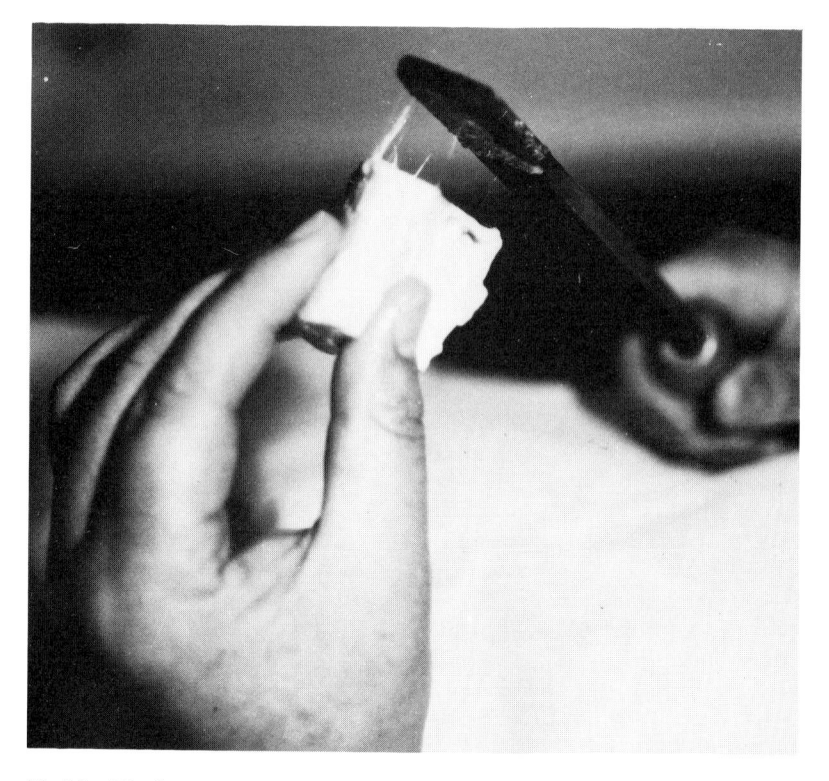

Fig 21 The hot iron test determines the state of the curd during the cheddaring stage. The firmness and length of strands shown are an indication of the level of acidity.

filling. The mercury-filled give a quick reading, and the alcohol-filled give a slower but clearer reading.

The temperature scales used for cheese can be in Celsius or Fahrenheit. The Celsius scale is more frequently used, although many cheesemakers prefer the Fahrenheit scale because there are more units of measurement per unit of heat. This enables the temperature to be finely adjusted.

Thermometers must be treated with care and cleaned after use, particularly if a protective cover is used because this can be a source of contamination. The accuracy should be checked monthly against a National Physical Laboratory thermometer or a certified thermometer. These are normally kept by all large laboratories, in teaching establishments, hospitals and Public Health Departments.

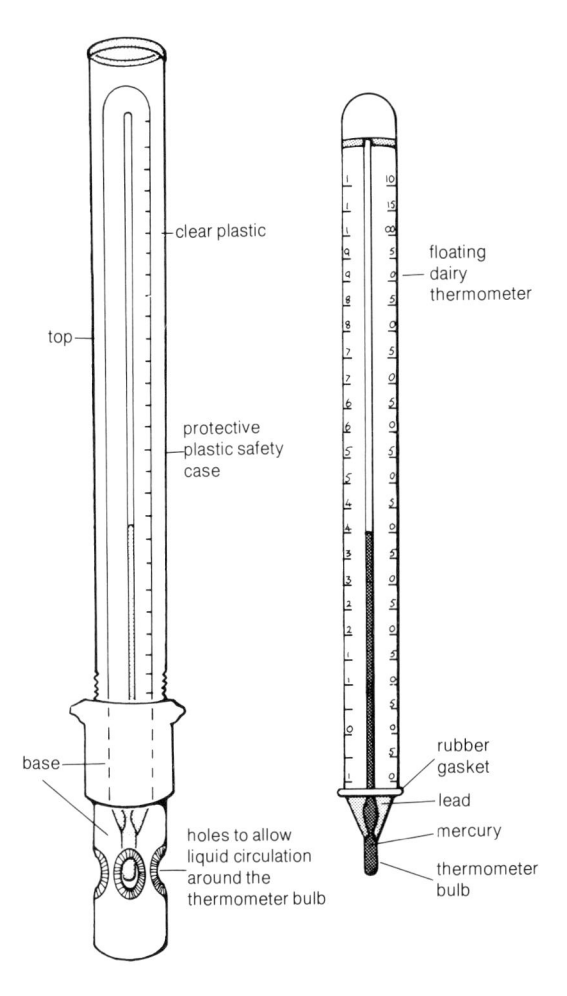

Fig 22 A floating dairy thermometer with a clear protective case to prevent breakages. The dairy thermometer is slid into the top part of the cover and the base of the cover is screwed on. The rubber gasket that has been positioned on to the thermometer base forms a watertight seal.

Measurement of Humidity

Traditionally bandaged or non-bandaged cheese needs a humid store, and this required humidity can be measured using a 'Wet and Dry

Bulb Hygrometer' (available from a horticultural supplier). This consists of two thermometers mounted side by side – the bulb of one is left exposed to the air, while the bulb of the other is covered with a piece of clean wet wick, kept constantly moist by dipping into a small vessel of water. The measurement of humidity is the difference between the two temperature readings, and is expressed as a percentage by using a conversion chart.

The apparatus must be positioned in a draughty area of the cheese store and the water vessel kept filled.

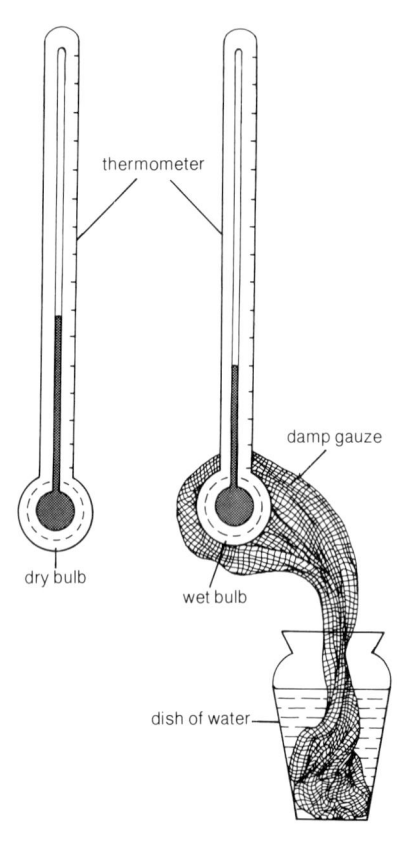

Fig 23   The hygrometer measures the temperature of a room with a dry bulb and a wet bulb. By using the table it is possible to use the temperature difference to determine the moisture content of the air. This is important in a cheese store that needs to be humid.

CONTROL TESTS AND CHEESE RECORDS

Records

There are three recording systems needed – the cheese process record, a record of the starters, and the cheese store record. If only a few cheeses are made the store record could be part of the process record.

An example of a cheese record sheet can be found below. The record must provide clear, accurate details of the day's make and, as

---

Cheese Record

Date _____ Vat Number _____

Variety of cheese _____
Cheesemaker _____
No. of starter _____
Volume of milk _____ Acidity of milk _____
Volume of starter _____ Acidity of starter _____

Time of starter addition _____
Volume of rennet _____
Time of rennet addition _____ Acidity at renneting _____
Time curd cut _____ Acidity at cutting _____
Time curd scalded _____
Temperature of scald _____ Acidity at scald _____
Time curd pitched _____ Acidity at pitching _____
Time whey off _____ Acidity at whey off _____
Time at first blocking _____ Acidity at blocking _____
Times of subsequent turns _____ Acidity at turns _____
    _____   _____
    _____   _____
    _____   _____

Time of milling _____ Acidity at milling _____
Quantity of salt _____
Time of salting _____
Time of moulding _____
Time and pressure at pressing _____
Number of cheese _____
Code of cheese _____
General remarks _____
_____
_____

Fig 24

61

long as it fulfils these requirements, it can take any form. An exercise book, carefully headed before making the cheese, is quite adequate. It is important to fill in the cheese record sheet during the day as the results become available.

The starter record is best kept in a small notebook and must record the history and activity of the starter and the quality of the cheese produced with it. Similarly, the cheese store record can be written down in a small notebook. It is important to record dates and cheese numbers, and to allow for grading remarks alongside.

7
The Cheese Recipe

Hard pressed cheese is made in a sequence of operations. These include ripening the milk with a culture of lactic streptococci and coagulating the milk by the addition of sufficient rennet to form a soft curd; cutting the curd into small evenly-sized pieces and cooking the curd to dry it; matting the curd after the whey is removed and finally milling, salting and shaping the curd ready for storage until matured.

The process has the effect of removing sufficient moisture from the curd to enable it to be matured. The period of time needed for maturing will vary, depending on the final moisture content of the cheese. Fig 25 shows how the milk constituents divide to form cheese.

The recipes for most hard cheeses are similar up to the stage of cutting. After this, slight differences in the treatment of the curd result in the cheese developing different textures and flavours. The recipe for Cheddar cheese is a good basic one and is used in the next chapters which describe the manufacturing process.

Cheese recipes can be written in many ways, but are usually found in the form given later in this chapter. The recipe is set out in order of the process operations and each stage is headed with a descriptive term. Many of the terms used are peculiar to cheesemaking and may be unusual to the layman.

The division of milk constituents during the manufacture of Cheddar cheese

Whey%		Milk%		Cheese%
83.10	←	Water 87.0	→	3.90
4.90	←	Lactose 5.1	→	0.20
0.30	←	Fat 4.0	→	3.70
0.10	←	Casein 2.5	→	2.40
0.65	←	Whey Protein 0.7	→	0.05
0.35	←	Salts 0.7	→	0.35
89.40	←	Total 100.0	→	10.60

Fig 25

THE CHEESE RECIPE

In an attempt to clarify the form of recipe used in this book, this chapter describes the process and some of the principles involved, using the headings that would appear in a cheese recipe.

Milk, Starter, Ripening and Annatto

Milk

Milk is used for cheesemaking because its chemical composition and physical properties enable certain biochemical changes to take place which can cause it to change state and form a curd. To ensure that the milk is safe for human consumption it should be produced from healthy animals, cooled as soon as it is produced, and pasteurised to destroy pathogenic bacteria.

Starter

Starter cultures are used in cheesemaking to produce lactic acid in the milk – this acid is needed to create a suitable environment for the action of the rennet enzyme. The acid also contributes to the flavour and texture formation of the cheese. Its production is brought about by growing a culture of lactic streptococci and adding it to warm milk at the rate of two per cent by volume. The bacteria produce lactic acid from the milk sugar lactose, by means of an enzyme lactase. The acid is produced throughout the process and the rate at which this occurs is controlled by the cheesemaker.

Ripening

Ripening is the period of time at the beginning of the process when the starter is left in the milk. During this period the lactic bacteria adjust to their new environment and begin to produce lactic acid. The ripening period is usually 30 minutes with two per cent starter.

Annatto

Annatto is added to the cheese milk to give added colour if the cheese variety demands it. It produces a pale creamy colour in the initial milk and, as the acidity increases, the colour intensifies, until at the end of the process it will have produced a deep orange colour, described in cheese terms as 'red'.

Renneting

The initial function of rennet is to coagulate the milk. After the curd is formed the rennet continues to change the state of the curd throughout the process. The rennet contains a proteolytic enzyme called rennin which acts upon the milk protein casein. Initially, the casein in the milk solution is denatured by the enzyme and forms a precipitate – this is curd formation. As the cheese process progresses the casein is further denatured and the curd changes from a soft curd to a firm pliable form with a completely different texture – this stage is called cheddaring. The rate of rennet action is controlled by the rate of acid development and by the temperature used in the process.

The enzyme will form a curd in warm, slightly acidic milk in about four minutes, so a four-minute limit is put on the stirring time after the rennet addition. This time is critical because stirring for longer will damage the curd as it forms.

The amount of rennet recommended in the standard recipe is for cows' milk of average composition. When other types of milk are used the rennet quantity must be carefully adjusted. The temperature is the same for all types of milk, as is the method of addition.

Cutting

When curd is set firmly it is cut into small particles of even size using sharp knife blades. The cutting releases the whey from the curd and the curd floats easily in the whey if kept stirred. The particle size required depends on the variety of cheese being made as it helps determine the moisture level in, and, therefore, the characteristics of the finished cheese. The loss of whey is much quicker from small particles (as with Cheddar) than from a larger cut (as used in Caerphilly).

The curd is very soft at this stage of the process and must be handled very gently to avoid further damage to the newly-cut surfaces of the curd. Harsh handling will cause loss of fat and other substances from the curd – this will reduce the cheese yield and have a detrimental effect on the quality of the cheese. Whatever the size of cut it is important to cut the curd into uniformly-sized particles so that they cook evenly.

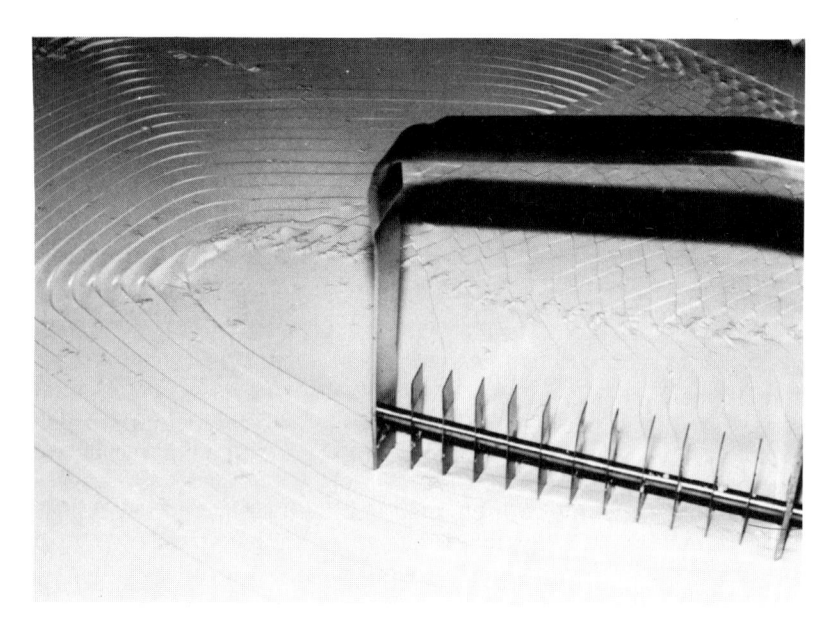

Fig 26 When the curd is firm enough it is cut using a knife or cheese blades. The curd pictured shows a firm curd that has been sliced cleanly.

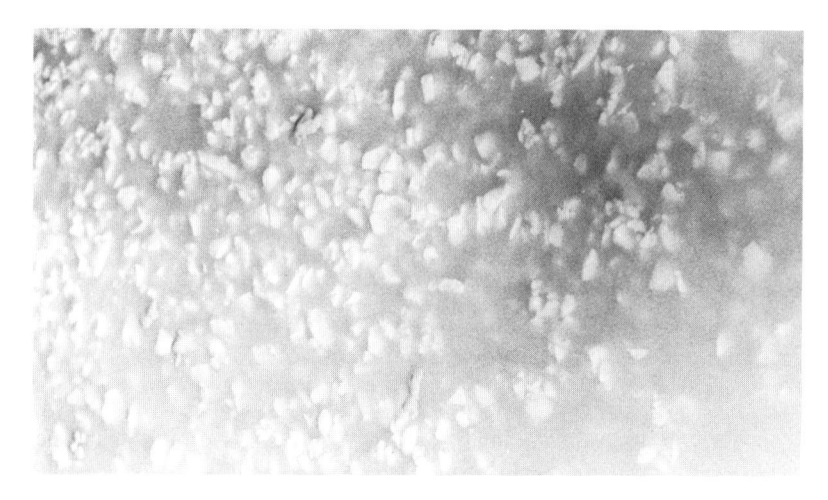

Fig 27 The cutting continues until the curd particles are even in size. During cutting, whey is released from the curd and the curd floats easily in the whey as long as it is stirred.

Scalding, Stirring and Pitching

Scalding

The curds and whey are gently heated over a period of time to a particular temperature. The time and temperature, in combination with the curd particle size and the rate of acid development, determine the final moisture content of the cheese. Scalding for Cheddar takes one hour and the curds and whey must be stirred continuously. The temperature is increased very gradually until the final temperature is reached. During this period the curd becomes firmer and begins to shrink due to increases in acidity and temperature which are encouraging further protein changes to be brought about by the rennet enzyme.

The scald temperature itself has the effect of slowing the rate of acid production because the temperature is not compatible with starter growth. The cheesemaker can make use of this factor to control the rate of acid development.

Stirring

Stirring takes place after the scald temperature has been reached. The curds are stirred for another hour. During this time the curd shrinks further and, as the temperature drops, the rate of acid production begins to increase. When it reaches the desired degree of firmness the curd is allowed to settle in the whey.

Pitching

Pitching is the term used to describe the settling of the curd – once it has done this it begins to knit together, and at this stage the whey can be removed.

Whey Off

The term 'whey off' describes the removal of the whey from the curd. The aim of the cheesemaking process is to remove the water from the milk in order to preserve the remainder of the milk as cheese. The situation at whey off is that the milk constituents which form the cheese are settled in the vat, and the whey (containing the water, lactose, whey proteins, water soluble vitamins and trace elements) can now be drained off. It is important that acid production is well established by this stage because once the whey has been taken off the main source of lactose has been removed. Slow cheeses may benefit from being left under the whey for a period of time before being drained.

Cheddaring

Once the whey has been removed, the curd is manipulated to encourage it to form a texture characteristic of the cheese.

Cheddar curd is blocked, and then turned and piled at intervals for about three hours. During this time the curd texture changes to become more homogeneous. This is as a result of rapid acid production by the lactic bacteria (the temperature will have dropped to a level which encourages its growth), increased activity of the rennet enzyme due to the increased acidity, and the gradual increases in pressure due to the continual turning and piling of the curd.

Once the texture has been changed and the acidity has reached the correct level the curd should be a rubbery dough-like substance with a silky feel. When torn, the layer of curd resembles chicken breast meat. At this stage it is ready to mill.

Milling and Salting

The cheddared curd has to be evenly salted (to help preserve the finished product), so it must be milled into small pieces to make this easier. The milling also has an effect on the texture of the cheese. An open-textured cheese, such as a Cheshire, is milled very finely and pressed lightly. A close-textured cheese is milled into larger pieces and pressed firmly.

Salting has several functions – it helps control the rate of acid development by slowing starter growth, it expels moisture from the curd (this increased rate of whey drainage helps dry the curd), it acts as a preservative by suppressing the growth of undesirable bacteria in the maturing cheese and it adds flavour to the cheese. In order to carry out all these functions effectively salt must be well mixed in, to ensure that it is evenly distributed throughout the cheese.

The salted curd must be allowed to drain and cool, with frequent mixing, until it is ready to be packed into moulds.

Fig 28   After whey off the curd is piled and cut into blocks. At this stage it has the appearance of scrambled eggs.

Fig 29   The curd blocks are turned, and, provided they are kept warm, the curd will begin to knit together.

Fig 30   If the blocks are large they are cut once more, as shown by students at Cannington College in Somerset.

Fig 31   The cut blocks are piled and as the acidity increases the whey drains out and the texture changes.

Moulding and Pressing

The salted curd is packed into lined moulds in order to be compressed and shaped. The moulded curd is then pressed together, to eliminate the excess whey, and to bring the salted curd together. The compressed curd will form a cheese which will be able to retain its shape unsupported during storage, provided it has been well made.

Presswork

This is the final treatment of the curd surface prior to storage. The aim of traditional presswork is to encourage the formation of a rind on the surface of the cheese, and to protect the surface from drying and cracking with the use of lard and calico bandages. Careful presswork protects the cheese during storage.

Storage and Maturing

The cheese is stored in a controlled environment, to enable all the variable factors of cheesemaking to come together in the maturing cheese. The storage must provide conditions that will encourage the steady balanced maturation of the cheese. Temperature and humidity must be at the correct level to ensure the desired rate of maturing and to prevent damage to the cheese surfaces due to any lack of moisture.

The length of storage is determined by the moisture level of the cheese. Moist cheeses, such as Caerphilly and Wensleydale, mature in weeks. Dry cheeses, such as Cheddar and Double Gloucester, mature in months.

Cheddar Cheese Recipe

Milk Raw milk, cooled immediately after production. Pasteurised milk 71.6°C (160.9°F) for 15 seconds. Initial milk acidity.

Starter Mixed culture of lactic organisms inoculated at temperature of 24–30°C (75–86°F) at a rate of $1\frac{1}{2}$–2 per cent.

Ripening Incubated for $\frac{1}{2}$–1 hour or until an acidity rise of 0.04 per cent LA from initial milk acidity.

THE CHEESE RECIPE

Annatto Use in winter milk. 15ml in 450 litres (½fl/oz in 100 gallons).

Renneting Milk temperature 30°C (86°F), rate of usage 25–30ml per 90 litres (1fl/oz/20 gallons). Dilute rennet 5–6 times with cold water. Add rennet throughout the vat and stir for four minutes; top stir for a further 1–2 minutes and leave the milk to set for 45–60 minutes.

Cutting When the curd is firm enough to cut (when it splits cleanly), the curd is cut lengthways and crossways with the horizontal and vertical knives and the acidity of the whey is tested. Cutting is then continued until the desired curd particle size is obtained – wheat grain to small pea size. Acidity drop of 0.04–0.05 per cent LA.

Scalding The curd and whey are stirred gently as the temperature is raised about 1°C (2°F) every 5 minutes, increasing to 1°C (2°F) every 3 minutes to scald temperature of 39–40°C (102–104°F). Scald is 60 minutes after cutting. Acidity at scald increased by 0.02–0.03 per cent LA.

Stirring After scald the curds and whey are stirred for a further 45–60 minutes.

Pitching When fit, the curds are pitched, acidity increase 0.04–0.05 per cent LA, i.e. 0.2–0.24 per cent LA.

Whey Off The whey is drained off, and the curd channelled; acidity 0.24–0.28 per cent LA.

Cheddaring The curd is cut and blocked, then turned and piled every 10–15 minutes until the acidity reaches 0.75–0.85 per cent LA, a rise of 0.05 per cent every 15 minutes.

Milling Mill the curd at 0.75–0.85 per cent LA.

Salting Curd salted at a rate of two per cent. Salt is well mixed in and the curds left to cool before moulding at 25.5°C (78°F).

Moulding Salted and transferred into lined moulds.

Pressing Pressure is applied gradually at first, then increased to 75kN/m^2 (15cwt) for 12–16 hours, then at 100kN/m^2 (20cwt) for the remaining days.

Presswork Routine – Traditional
Day 1   Put to press in rough cloth and increase the pressure gradually to 100kN/m^2 (20cwt).
Day 2   Remove from press, carefully remove cloth, bathe in hot water at 60°C (140°F) for 30 seconds. Return to press in a clean, dry, smooth cloth, and repress.
Day 3   Remove from press, lard and bandage cheese; re-press.
Day 4   Remove from press, lard and bandage and leave to dry on clean board; label with number or code and remove to cheese store.

Maturing Cheese store humidity 80–90 per cent, temperature 10–13°C (50–55.5°F). Turn cheese daily for one month and weekly thereafter, until mature.

This is a typical cheese recipe which gives details in an abbreviated form with varing amounts of information. The following two chapters will use this recipe as a base to explain and describe in detail the cheese process.

8
Practical Cheesemaking

The practical cheesemaking described in this chapter will apply to all quantities of milk used unless otherwise stated. Variations in the process which need to be made to allow for the differing compositions of the milk will be included where necessary at each stage of the process. The cheesemaking process is described in this chapter in a detailed, practical manner and emphasis is placed on the interpretation and application of the recipe with the areas of importance being stressed.

The process details are given assuming that all the factors involved allow the cheese to follow the plan laid out in the recipe. In practice this rarely happens and adjustments have to be made; these adjustments and remedial procedures are to found in the following chapter.

Preparation

A bulk starter of mixed culture will have been prepared the previous day, the end of its incubation period having been planned to coincide with the proposed time of starting the cheese process.

The cheesemaker must wear clean overalls and hat, and hands and arms must be well scrubbed. The apparatus for the acidity test is prepared and all equipment needed is checked and made ready.

Note All equipment must be clean and scalded with boiling water, or rinsed in a suitable solution of hypochlorite before use.

Stage One – Ripening

This stage includes milk, starter, ripening and annatto, and is the part of the process where the lactic streptococci are inoculated into the milk and their steady growth established. It is also the step at which cheese colouring is added to the milk.

Equipment, reagents and services required for cheesemaking

Equipment	Reagents
Acidimeter	Starter
Vat	Rennet
Plunger or stirrer	Salt
Thermometer	Lard
Cheese knives or knife	Annatto
Sieve	
Bucket	**Services**
Curd knife	Hot and cold water
Board	Boiling water
Small weight	Steam
Cloths and bandages	
Mill or knife	Detergent
Moulds	Sodium Hypochlorite
Press	

Fig 32

The vat, stirrer and thermometer are scalded – special attention must be paid to scalding the tap area of the vat.

The milk, raw or pasteurised, is transferred to the vat as smoothly as possible to prevent damage to the milk fat. The milk is then heated to a temperature of 26°C (78.8°F) and an acidity test is carried out. The expected acidity of milk is 0.14 per cent LA (Channel Island 0.18 per cent LA, goat 0.19 per cent LA, sheep 0.22 per cent LA).

The starter is then checked. The lid of the starter can is lifted and the starter sniffed immediately to detect 'off' flavours or faults. It is then well mixed and an acidity test carried out – the result should be approximately 0.85 per cent LA.

The quantity of starter is calculated and added to the milk (2 per cent by volume). The starter is well stirred in and the milk left to ripen for 30 minutes. It is advisable to cover the vat with a sheet to help prevent contamination.

The quantity of annatto required is calculated and added to the milk at least 15 minutes before renneting. It must be well stirred in to ensure that the finished cheese will have a uniform colour.

Stage Two – Renneting

The addition of rennet must be undertaken with care – damage to the curd structure can easily occur at this stage and ruin the cheese. The rennet quantity is calculated and diluted with six times its volume of *cold* water. The milk is uncovered, stirred well and an acidity test is carried out. The expected acidity will be an increase of 0.05 per cent LA from the initial milk acidity to approximately 0.19 per cent LA. The milk is warmed to 30°C (86°F) and the rennet solution is added. Pour the rennet evenly throughout the vat and stir evenly for four minutes; at four minutes, stop stirring and remove the stirrer. Top stir for a further two minutes. This is done by gently stirring the top 1–2cm (½–1in) of the milk surface with a spoon or the base of the stirrer. The reason for this operation is to prevent the fat globules from rising to the surface – in this way they are trapped in the curd and this will improve the texture and yield of the cheese. After top stirring the vat is re-covered and the milk left to coagulate for 45–60 minutes.

It is particularly important not to let the milk become chilled during

Fig 33 Rennet is used in the same proportion for all volumes according to the type of milk. It is advisable to dilute it with cold water to ensure even mixing.

Fig 34   Once the rennet is added, stirring is continued for four minutes only.

Fig 35   The renneted milk is left to set once the four minutes stirring time is completed.

this time as this will affect the formation of the cheese curd. With sheep's milk there is also a risk of over-stirring if the rennet quantity is too high. If in doubt, reduce the stirring time to three minutes.

Stage Three – Cutting

Cutting must be carried out carefully. The curd at this stage is very soft, and serious damage is caused if it is roughly handled. The aim of cutting is to produce curd particles which are uniform in size – this enables the subsequent cooking to be done evenly. The curd is checked for fitness to cut 45 minutes after the rennet is added. This can be done as follows:

A clean, dry hand is placed palm upwards, on the surface of the curd and pressure is applied gently downwards, the fitness of the curd is ascertained, and the hand then lifted away from the curd. Curd which is ready to cut will feel resilient and will not stick to the back of the hand. Curd which is not ready feels soft and sticks to the hand.

The curd knives or knife are scalded ready and the curd is cut.

Cheese in a Vat

Curd is cut using American curd knives, the vertical blade knife first. The knife is inserted vertically to the bottom of the curd at one end of the vat, and, held upright, it is moved the length of the vat sufficiently fast to allow the curd to be cut without it moving away in front of the knife. Further cuts are made lengthways down the vat, slightly overlapping the previous cutline. When the lengthways cuts are complete, the curd is cut at right angles across the vat, using the same knife and the same technique.

The knife is removed and an acidity test carried out on the whey. The acidity will have dropped 0.05 per cent to approximately 0.14 per cent LA.

The horizontal knife is then used in a similar manner, lengthways and crossways in the vat, overlapping each run slightly. It may be necessary to increase the speed of cutting because the curd will tend to sink at this stage. The curd is further cut into small evenly-sized pieces, using a vertical knife randomly across the vat. Check the vat corners for uncut curd. Curd size is determined by the acidity, and in normal conditions this will be wheat grain size. The curd is stirred very gently as soon as cutting is complete.

Fig 36   The curd is cut into even-sized pieces according to the variety of cheese. The curd shown is being cut with an old diagonal-bladed knife.

Fig 37   The curd can be cut using a vertical blade knife and a horizontal blade knife. The vertical blade knife is shown in use in a teaching dairy.

Cheese in a Small Container

This can be cut using a sharp knife or spatula. With the knife in a vertical position lines are cut across the bowl to the depth of the curd – these lines should be 1.25cm ($\frac{1}{2}$in) apart. The curd is cut again, at right angles to the first cutlines and using the same method (*see* Fig 38). Remove the knives and leave the curd for two or three

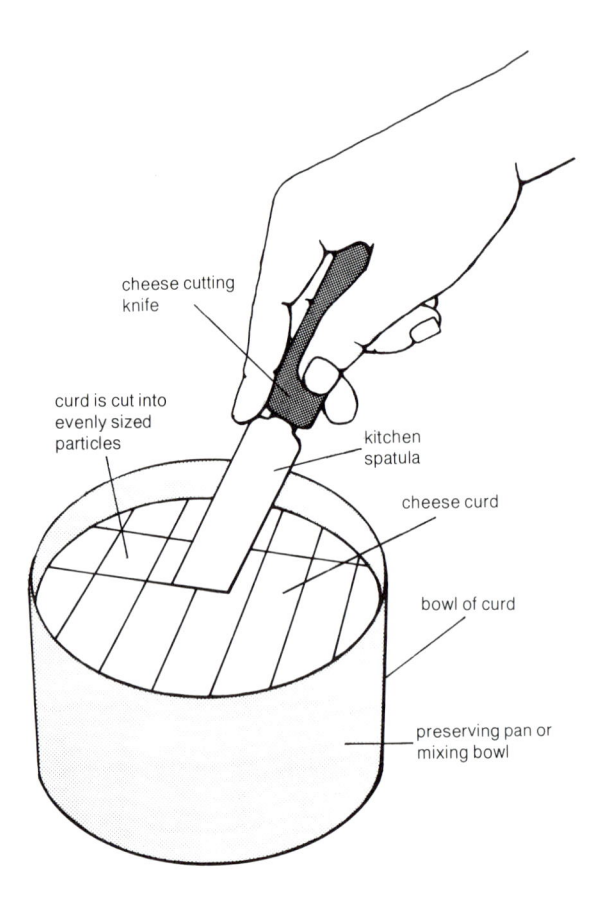

Fig 38 Curd in a small container can be cut with a kitchen knife or spatula. The curd is cut across the bowl first and then again at right angles to the first cut. It can be cut randomly thereafter until even and of the required size.

minutes to allow some of the whey to accumulate. Take an acidity test – the acidity will have dropped 0.05 per cent LA to around 0.14 per cent LA.

The curd is now cut further (using the knife at a diagonal angle) progressively across, and then randomly, until the pieces are uniform, the size of wheat grain. The curds and whey are then stirred very gently – the stirring will be continuous up to the stage of pitching.

Sheep's Milk

This may have to be treated slightly differently. Firstly, the setting time may be less than 45 minutes; and secondly, the curd formed will not produce as much whey as other milks because of the high solids content. In order to stir the curds and whey without damaging the curd an amount of clean warm water can be added. The quantity of water must be sufficient to enable the curds to float freely and it is added at a temperature of 30°C (86°F).

Stage Four – Scalding, Stirring and Pitching

Scalding and stirring influences the final moisture content of the cheese, and must be carried out strictly as described in the recipe in order to ensure the correct moisture level in the finished cheese.

Scalding

This stage of the process involves the gentle application of heat to the curds and whey in order to raise the temperature from 30°C (86°F) to about 40°C (104°F). This can be carried out by heating in the vat jacket using hot water or steam. An improvised vat will be heated using hot water in the outer container.

The timing of scalding is calculated from the beginning of the curd cutting and takes exactly one hour:

For the first 15 minutes the curd is not heated at all.
For the next 30 minutes the temperature is raised by one degree for every five minutes to 36°C (96.8°F).
For the last 15 minutes the temperature is raised by one degree every three and a half to four minutes to achieve a temperature of 40°C (104°F) at one hour.
An acidity test is taken – this should show an increase of 0.02 per cent LA to about 0.16 per cent LA.

Fig 39 Stirring during cooking is carried out carefully to avoid smashing the curd. The thermometer shown is covered with a protective plastic sheath to prevent breakages.

Stirring

Once the scald temperature has been achieved the curd is stirred for a further hour until fit to pitch. During this hour the curd shrinks – this shrinkage is due to the combined effect of rennet and heat. After about an hour the curd will be dry enough to pitch.

At this stage the curd will feel 'shotty' – in other words, when the hand is run through the whey the curd particles feel hard. This is difficult to determine. An acidity test is taken, and this should show a rise of 0.02–0.03 per cent LA to approximately 0.19 per cent LA.

Pitching

This is the settling out of the curd in readiness to draw the whey. Stirring is stopped and the curd is allowed to settle on the bottom of the vat. The time that the curd is left pitching varies with the quantity of milk. A small amount in a bowl should be left for 30 minutes, while 180 litres (40 gallons) in a vat needs about 10–15 minutes. This period enables the acidity to build up before the whey is lost.

Stage Five – Whey Off, Channelling and Blocking

Whey Off and Channelling

The whey is drawn off in order to remove most of the water and lactose, ready to cheddar the curd. The loss of the lactose reduces starter activity and helps control acid production.

Cheese in the Vat

The curd is pushed gently away from the vat tap and the whey is drained off through a sieve. The curd caught in the sieve is returned to the vat and the whey is collected and stored ready for disposal.

When the whey has been removed a channel is made down the centre of the vat to enable further whey drainage. An acidity test is carried out and there should be a rise of 0.05 per cent LA to 0.24 per cent LA.

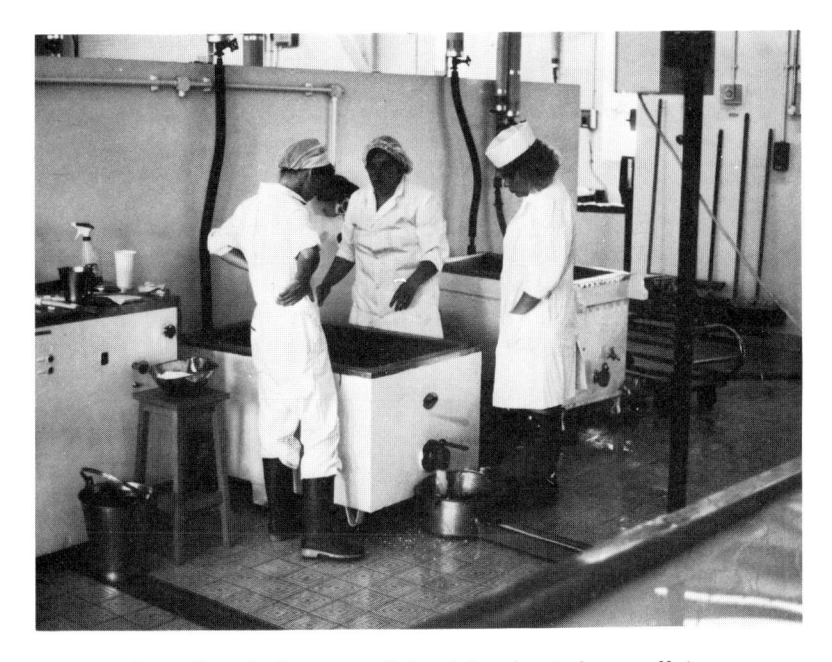

Fig 40 The curds and whey are settled and the whey is drawn off via the vat tap into a sieve and transferred to a storage area.

Fig 41 Once the whey is off the curd is channelled to drain the whey further; this can be done with a knife or clean hands.

Cheese in a Small Container

Drain the whey off the curd and transfer the curd to the centre of a freshly scalded cloth. Draw the four corners of the cloth up around the curd. Hold three and take the fourth corner several times around the base of them. This forms a Stilton bundle, as in Fig 42.

The bundle is put on a board over a bowl or the sink and an acidity test carried out, as above.

Blocking the Curd

The curd is cut into 20cm (8in) wide blocks and left for five minutes. The blocks are then turned over ready for cheddaring. The curd should be kept warm by using clean dry cotton cloths, especially in winter.

PRACTICAL CHEESEMAKING

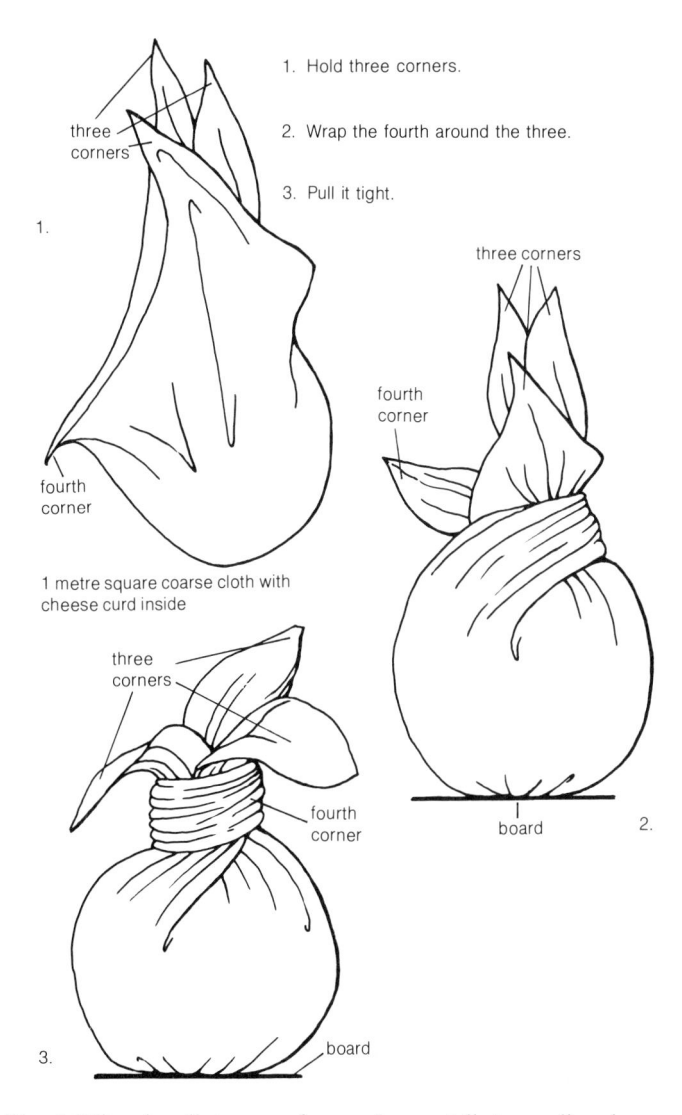

1. Hold three corners.
2. Wrap the fourth around the three.
3. Pull it tight.

Fig 42 A Stilton bundle is very often used, especially in small-scale cheesemaking. The curd is placed in the centre of a scalded cloth and three of the cloth corners are pulled up. The fourth corner is wrapped around the other three, and continued round until used up. The result is a tight secure bundle which retains its heat.

Fig 43 The curds and whey are settled and the whey is poured from the kitchen vat through a sieve into a bucket.

Fig 44 A small quantity of curd is transferred to a cloth after whey off and tied in a Stilton bundle to keep it warm.

Fig 45 Large volumes of curd fuse together quickly due to the weight and heat retention.

Fig 46   Small volumes of curd tend to become chilled and may remain crumbly, as shown.

Stage Six – Cheddaring

Cheddaring is the stage at which the curd texture is formed by the combination of acidity, warmth and weight.

Cheese in a Vat

The blocks are now turned and piled every 15 minutes, and each time they are turned they are piled a little higher:

First turn – curd turned and piled two high; acidity 0.29 per cent LA.
Second turn – curd turned and piled three high; acidity 0.34 per cent LA.
Third turn – curd turned and piled four high; acidity 0.39 per cent LA.

There is a limit to the number of times it is physically possible to pile the blocks – when this stage is reached, turning should simply be continued. The curd must be kept warm and if there is only a small amount it may be necessary to consolidate it with the help of a weight.

Fig 47 Small volumes of curd can be cut into fillets and reweighted once they have knitted together. This will encourage whey drainage.

Fig 48 Small volumes of curd should be covered with a cloth and weighted using a board and small weight. This helps retain heat and encourages curd texturing.

This can be done by covering the curd with a cloth and placing a weight on a board on top of the cheese.

During Cheddaring

The acidity is tested every 15 minutes and there should be a rise of 0.05 per cent LA each time. When the acidity is 0.75–0.85 per cent LA the curd should be ready to mill. With small amounts of curd there is often not enough whey available to test and the hot iron test or a considered guess will be needed to determine when the curd is ready to mill. When cheddared, the curd looks smooth and silky, and when a strip is torn it resembles cooked chicken breast meat.

Cheese in a Small Container

The cheese wrapped in the bundle is cut in two and rebundled. The two blocks of curd are then turned and rebundled every 15 minutes for two and a half hours. After one hour put a board and weight on the cheese to help consolidate it. Ensure that the curd does not become chilled during this process. It will not be possible to take acidity tests after whey off, so cheddaring time has to be estimated.

Fig 49   Kitchen volumes of curd must be wrapped in a cloth and weighted to encourage curd texturing.

Stage Seven – Milling and Moulding

Milling

Milling is carried out when the curd is cheddared and its purpose is to reduce the curd to uniformly-sized pieces of about 3–5cm^2 (1½–2in^2) to enable the salt to be distributed evenly through it.

The cheddared curd is cut into small pieces using a mill or a knife; when this has been completed the salt is added at the rate of two per cent by weight, and mixed in very thoroughly.

The curd is allowed to cool to 26°C (78.8°F) and is then moulded.

Moulding

The moulds are lined with suitable cloths of cheese-grey or nylon net. They are then filled with curd – several handfuls are put in at a time and then pressed down with the fist, until the mould is full. The cloths are then pulled up from the side of the cheese with one hand while the other presses firmly on the top. The followers are placed into position and the cheese put to press.

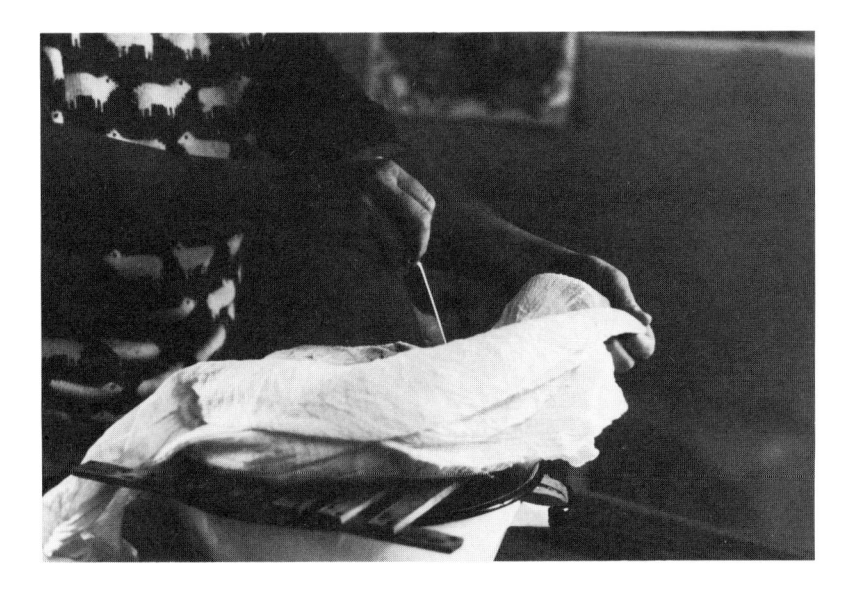

Fig 50 Kitchen volumes of curd are cut into fillets, re-wrapped and weighted.

Fig 51   The curd is cut into cubes of uniform size and salt is spread on evenly and mixed in well.

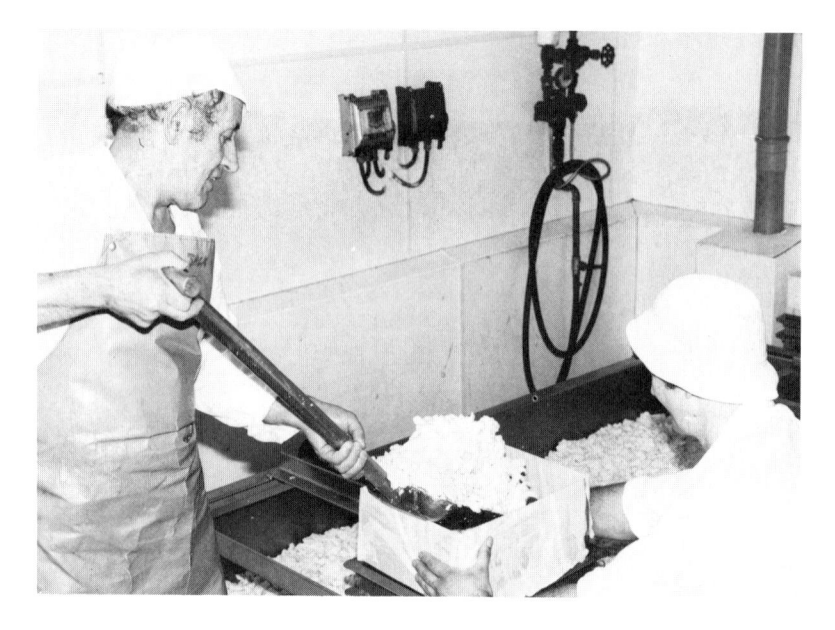

Fig 52   The salted mixed curd is transferred into lined cheese moulds when sufficiently cool.

Fig 53 The freshly-scalded mould is lined with a clean cloth and placed on a mat and board in a tray.

Figs 54 and 55 The salted curd is placed into the lined mould and pressed down as the mould is filled. When the mould is filled the cloth is pulled across the cheese surface and the wooden follower positioned in the top.

Fig 56   The pressure is applied to the fresh cheese very gradually and the whey is collected in the base tray.

Stage Eight – Pressing

The cheese is pressed for three days to consolidate the curd. The initial pressing is important. The pressure must be applied slowly for the first two hours up to 49.9kN/m^2 (10cwt) and can then be increased to 100kN/m^2 (20cwt) for the remainder of the time. This gradual increase prevents any whey becoming trapped in the curd.

Improvised presses must be used in a similar way – low pressure initially, followed by a higher pressure – and their effectiveness will have to be judged on the resulting cheese.

9
Remedies and Alternative Actions

The previous chapter gives an account of a cheesemake going exactly according to plan. This rarely happens and adjustments to the process often have to be made. Inexperience or mismanagement may cause problems and it is important to know how to rectify mistakes. This chapter goes through the procedure once more, describing alternative and remedial action that can be taken, if found to be necessary.

Preparation

Starters

These are prepared the previous day and must only be used if they have a clean, sharp, acid smell. If any 'off' flavours have developed during incubation the starter should not be used, and the starter technique must be checked. If the starter acidity is slightly less than 0.85 per cent LA, but smells clean and slightly acid, the start of the cheesemaking should be delayed and the temperature of the starter verified, as it might have become chilled during incubation. The culture can be warmed if necessary and incubation continued until the acidity has reached 0.85 per cent LA, the cheesemaking may then proceed.

Starter which is contaminated or of very low acidity must be deemed failed.

An alternative source of culture is DVI culture and this can be used following the DVI Cheddar recipe (*see* Chapter 12).

Milk Source

The source of milk may be fresh or frozen. Fresh milk must be cooled immediately it is produced and it is recommended that all milk should be pasteurised. Frozen milk must be thawed out completely and this must be done carefully in order to prevent contamination – defrosting milk in the refrigerator tends to take too long, whilst thawing milk at ambient temperature may lead to bacterial faults because some of the thawing milk might become warm enough for bacterial growth. The frosted milk should be thawed in a covered container which is then placed in a large water-filled bowl. It could also be placed in a covered cheese vat, and the vat jacket filled with cold water. The water jacket in each case ensures even thawing and keeps cool the liquid milk which has thawed.

Channel Island Milk

Channel Island milk must have some of the cream removed before it is used for cheese, and without laboratory facilities it is rather difficult to judge how much needs to be discarded. Removing half to three-quarters of the fat from the evening milk after it has settled overnight, will give a fat percentage of approximately the correct amount when it is mixed with the morning's milk. Full fat Channel Island milk makes a rather greasy cheese because the large globules of fat found in the cream are not easily entrapped in the curd. This problem does not occur with goats' and sheep's milk because, although the fat percentage in the milk is high, the globules are small and thus are easily trapped in the curd.

Milk Acidity

The initial acidity of the various milks will range from 0.12 – 0.22 per cent LA. The milks with a high percentage of solids have more natural acidity and a higher reading. The acidities measured during the cheese process will therefore have a correspondingly high reading. If milk has an acidity reading which is much higher than normal it may be due to poor quality or to the acidity test being incorrectly carried out.

Stage One – Ripening

Addition of Starter

To ensure a steady rate of acid production the rates normally used are two per cent starter incubated in the cheese milk for 30 minutes, or one and a half per cent starter for 45 minutes. These times may be increased by 15 minutes when the weather is cold.

These rates and times allow the starter bacteria to become established before the rennet is added. Rates below one and a half per cent would need a longer ripening time and leave the starter culture vulnerable to a phage attack for a longer period. Rates above two per cent result in a cheese which is acid in flavour and has a short, coarse texture. The increased rate of acid production would be difficult to control during the process.

A short ripening period will result in a weak-bodied, under-acid cheese, due to the starter culture not becoming sufficiently well established before renneting. If the ripening period is too short, more time must be taken over all subsequent operations and the scald temperature should be reduced by $1\,°C$ ($2\,°F$).

A long ripening period results in acid being produced too quickly throughout the cheese process and the finished cheese is over-acid and short in texture. If the ripening period given is too long, all subsequent operations must be speeded up slightly, the scald temperature increased by $\frac{1}{2}-1\,°C$ ($1-2\,°F$) and turning during cheddaring carried out every 10 minutes instead of every 15 minutes.

The aim of the starter culture is to keep acid production in line with the physical changes during the cheesemake.

Annatto is added during ripening and this must be done at least 15 minutes before renneting to ensure that it is well mixed in. If the addition is late it must be mixed in vigorously to prevent a streaky coloured cheese. This may hinder starter growth as air may become incorporated. Colouring cannot be added after the ripening stage.

Stage Two – Renneting

The combination of rennet, heat and acidity will produce a curd in fresh milk. The procedure for adding rennet to milk must be carried out carefully.

Stirring

A stirring time that is too short results in uneven setting, and cutting must be delayed until the softest areas are firm.

Too long a stirring time is a disaster – the curd structure starts to form after four minutes and over-stirring breaks the structure. There is little that can be done to reclaim the cheese if a curd does not form. If the resulting curd sets, many of the milk constituents will not have been trapped and will be lost in the whey – this will lead to a reduced yield of cheese. If the curd is handled very carefully to prevent further damage, the cheese may be saved. Over-stirred curd looks as if it contains small curd particles.

Factors Affecting Stirring Time   Ripened milk with acidities of 0.2 per cent LA and above will tend to react with rennet very quickly, and it may, therefore, be prudent to reduce the stirring time if acidities are at this level.

The optimum temperature for rennet action is 40°C (104°F). If milk temperatures are above 30°C (86°F) then the rate of rennet action will increase accordingly and the stirring time must be reduced. Milk temperatures below 30°C (86°F) do not have a detrimental effect on the stirring time but the curd will take longer than normal to set.

Setting

A low temperature or low acidity will extend the setting time and cutting will have to be delayed until the curd is fit.

A high temperature or acidity will shorten the setting time and cutting will take place earlier.

Stage Three – Cutting

Curd is cut to allow the whey to escape and to make the curd particles small so that they can be cooked evenly. The size of the curd particles determines the final moisture content of the cheese.

The decision to cut the curd is a critical one and can be difficult if, for instance, the curd surface has become chilled. The following methods of how to determine curd firmness may be used as alternatives to the method given in Chapter 8.

Method 1.   Making sure it is clean, push the hand diagonally into the curd to a depth of 8cm (3in). Lift the fingers gently and the curd will split – if the curd is ready to cut then it will do this cleanly. This is a useful method if the curd surface is chilled.

Method 2.   Place a clean, dry hand palm upwards on the curd surface, at the side of the vat, making sure that the hand is parallel with the edge. Press down gently and at the same time pull the curd away from the vat side. Curd that is ready will feel resilient and will pull cleanly away from the vat side.

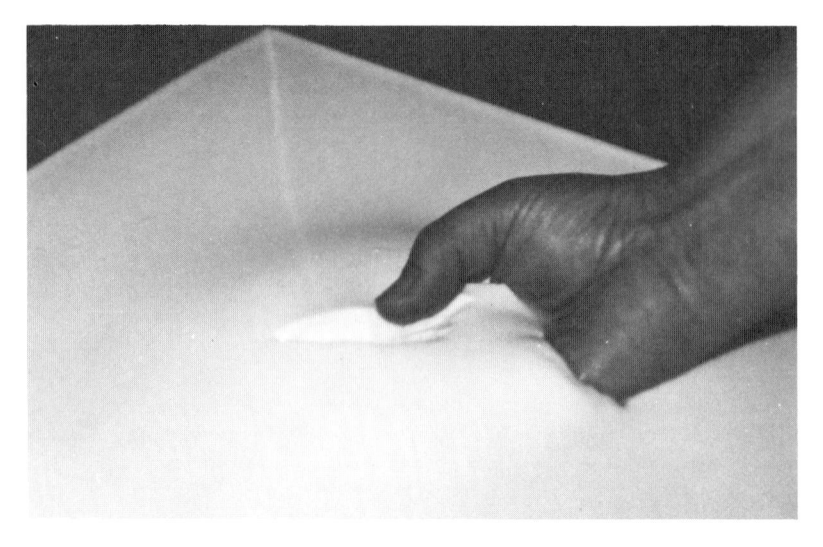

Fig 57 When the curd is set sufficiently it is possible to split the curd cleanly without tearing, by lifting it gently with the fingers as shown.

If the curd is cut too soon (when it is soft) it will be damaged, resulting in the loss of fat, and a reduced yield of cheese. The softness of the curd will be noticed when the cutting begins, in which case the procedure should be stopped and the curd left to set for a while longer. If the softness is not noticed until a later stage (very cloudy whey will indicate that the curd is soft), the curd should be left undisturbed for 10-15 minutes after cutting. This will allow the surfaces of the curd particles to 'heal', and should be done before starting the scald.

If the curd is cut rather late it tends to be difficult to produce uniformly-sized particles and this results in uneven cooking during scalding. The cheese may have a streaky appearance due to the varying moisture levels of the curd. The acidity at cutting will also be at a higher level than normal and it may be necessary to increase the scald temperatures by $\frac{1}{2}-1\,^{\circ}\text{C}$ ($1-2\,^{\circ}\text{F}$).

Stage Four – Scalding, Stirring and Pitching

Scalding

The curd is scalded to expel whey from the particles, control acid development and to produce the characteristics of the cheese variety.

Scalding must take place very gradually at first, to ensure slow, even expulsion of whey from the curd particles. If scalding is too fast the acid development will tend to be slow and it may be necessary to reduce the temperature by $\frac{1}{2}-1\,°C$ $(1-2\,°F)$. A fast scald can also cause 'case hardening' – the curd surface develops a rubbery skin and the whey cannot escape. The resulting cheese is very acid and weak in body, with a tendency to leak whey during storage.

If scalding takes place too slowly the rate of acid development will rise too quickly and may need to be controlled by increasing the scald temperatures by $\frac{1}{2}-1\,°C$ $(1-2\,°F)$.

If the scald temperature is higher than recommended the starter culture may be destroyed and the subsequent operations will have to be carried out very slowly. It may be helpful to leave the curd under the whey for a long period of time (one to two hours). Whey contains

Fig 58   It is very important to check the rate of temperature rise during the scalding stage of the process.

the lactose needed by the bacteria, and the longer the whey is retained the more chance there is of salvaging some of the starter bacteria. Once the whey is drained the blocks should be left large and kept warm.

A scald temperature that is too low will result in the starter bacteria producing acid at too fast a rate later in the process. It will be necessary to draw the whey off quickly after pitching and to cut the curd blocks small. The blocks may need to be turned more frequently than normal.

Stirring

Curd size, scald temperatures and stirring time all have a considerable effect on the characteristics of the finished cheese.

Insufficient stirring after scald will result in a higher curd moisture than normal, and the acidity during cheddaring may be a little slow to start with. The curd blocks should be cut larger and turned less frequently until the acidity builds up – then they should be turned as normal.

Too much stirring results in a crumbly curd which is difficult to handle once the whey is off – it does not knit together easily. This is not particularly detrimental to the cheese and the curd texture can be improved by using a weight on the curd piles during cheddaring.

Stage Five – Whey Off, Channelling and Blocking

The whey is drawn off to remove the water and to control the acid development. It is a rich source of lactose and acid development will be rapid whilst it is still in the vat. This factor can be used to encourage acid development in a slow cheese – the curd is left for longer under the whey after pitching.

If the whey is removed too fast the subsequent cheddaring may progress rather slowly. If the whey is removed too slowly the acid development will tend to be fast during cheddaring.

Channelling and Blocking

Once the whey is drawn off the acid development of the curd has to be determined.

Acidity in excess of 0.05 per cent LA in every 15 minutes requires the curd blocks to be cut small and turned frequently to help expel the whey. At this rate of acid development, if cutting and turning are too slow the curd will not be sufficiently cheddared when it reaches the milling acidity. This is not very serious, but may lead to a texture in the finished cheese that is less than perfect.

Acidity of less than 0.05 per cent LA every 15 minutes requires the curd blocks to be large and turned less frequently. By restricting whey drainage thus, acid production is encouraged.

Stage Six – Cheddaring

Curd is cheddared to encourage texture formation. The block size and frequency of turning controls the rate at which the acid develops at this stage.

Turning that is too rapid results in a slowing-down of acid development, caused by the loss of whey. This is not detrimental but will lengthen the whole process.

Too slow turning results in acid developing too fast, and acid cheese if the curd is left too long before milling. An increase in the salting rate will help slow down a very fast cheese.

Stage Seven – Milling and Moulding

Milling

The curd is milled to reduce it into evenly-sized particles ready for salting.

If the curd is milled too early, delay the salting and stir the curd every 15 minutes until the acid level required is reached. It is then salted as normal. Alternatively, the salt quantity could be lessened a little to allow for the slightly reduced acidity.

If the curd is milled late, extra salt could be added to help control the acid development. Rapid cooling would also slow down acid development.

Salting

Salt is added to retard acid production, to improve the rate of whey drainage by shrinking the curd and to enhance the flavour of the cheese. Salt will also suppress the growth of undesirable organisms in the cheese.

Low salt concentration leads to an acid cheese with a high moisture

REMEDIES AND ALTERNATIVE ACTIONS

content and bitter flavours. High salt concentration will retard ripening and the cheese will have a low moisture level, giving a dry, crumbly texture.

Mixing after salting is important to ensure uniform distribution of salt and even drainage and ripening. Uneven salting causes uneven maturing and may produce a mottled appearance in the finished cheese.

Moulding

Salted, cooled curd is moulded to allow it to be consolidated – this produces cheese of uniform shape and size.

The curd temperature should not be higher than 26°C (78.8°F) at the time of moulding.

A moulding temperature that is too high will result in fat and curd loss – these will be expelled through the drainage holes of the mould if they are warm and soft.

If the moulding temperature is too low, the curd will not knit together during pressing. This may be rectified when the cheese is bathed on the second day of presswork.

Stage Eight – Pressing

The curd must be pressed slowly at first to ensure that whey is not trapped within the cheese.

Pressure applied too fast results in whey being trapped and the cheese will leak whey as it matures.

Pressure applied too slowly results in curd that may become chilled before it is consolidated. This prevents it from knitting together and may result in a cheese with an open texture that is subject to mould contamination. Bathing on the second day of pressing may help correct this fault.

10
Presswork and Storage

Good organisation and work routines in the pressroom and the cheese store will ensure that well-made cheese is still in excellent condition by the time it is mature. The result of poor pressroom routine or bad storage conditions is a cheese that will develop faults whilst maturing, thus reducing its value.

Presswork Routines

The first day of pressing usually refers to the same day that the cheese is made and put into press (*see* Chapter 8). The routine for the second, third and fourth days is as follows:

Day Two

Equipment needed for the second day of presswork: a strong table or bench, a small bowl, a sharp knife, a smooth calico cloth and a bucket of water at 50–60°C (122–140°F).

The cheese is removed from the press and placed on the bench. The wooden followers are carefully removed and the surplus curd above the metal follower is cut away with a sharp knife and placed in the bowl. The metal follower is then removed and the cloth gently pulled away from the cheese, taking care not to tear the surface of the curd.

The cheese mould is turned upside-down, lifted, and banged down on the bench to dislodge the cheese – this enables it to be pulled from the mould. The cheese may need to be banged very hard, in which case a board on the floor may be a more suitable place to dislodge it.

With the cheese dislodged, the mould is turned upright and, using the cloth from the side of the cheese (not the cloth from the top) the cheese is pulled gently from its mould and placed on the bench. During pressing the cheese curd squeezes through the cloth, and this curd must be carefully scraped with a knife and removed before the cloth can be pulled away from the cheese. If this excess is not removed, curd will be torn away from the cheese leaving a damaged

PRESSWORK AND STORAGE

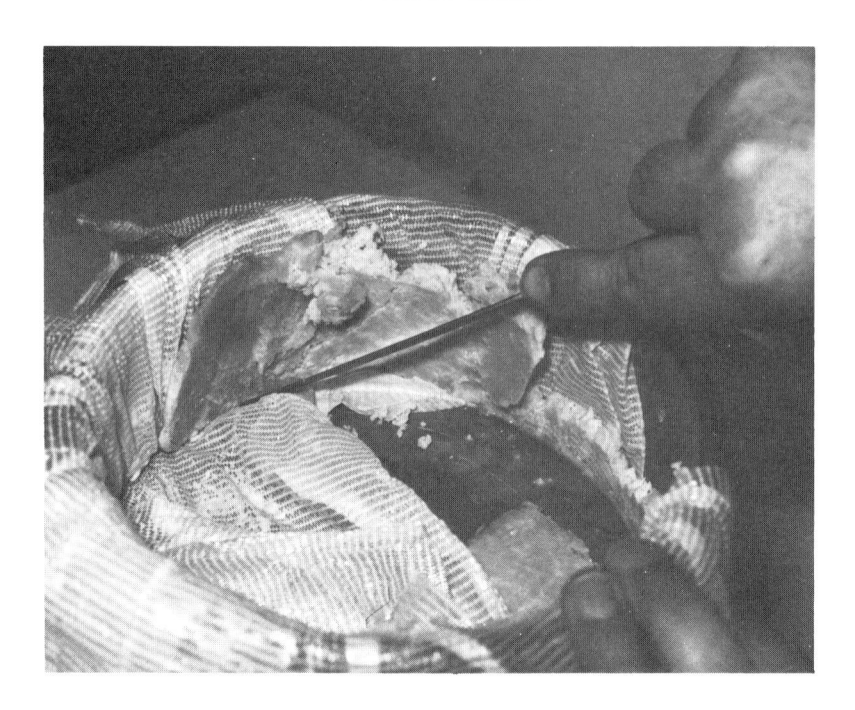

Fig 59 After the first period of pressing, the surplus curd above the follower must be carefully removed with a knife.

Fig 60 Once the curd is removed from the mould any surplus is trimmed to give a smooth edge.

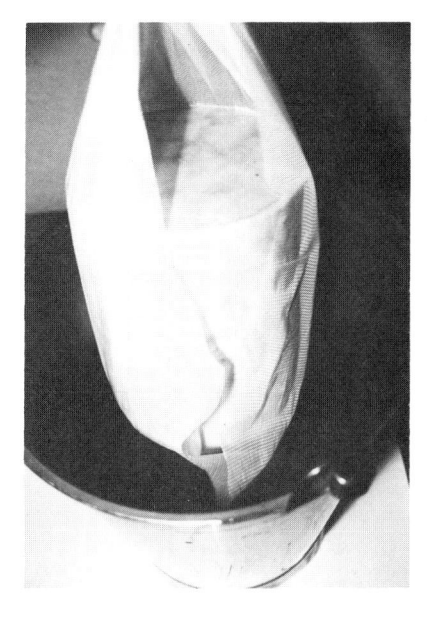

Fig 61 The trimmed cheese is immersed in hot water to remove surface fat and to encourage rind formation.

surface that is likely to become contaminated by mould growth in the store.

Curd is scraped away from the top surface and edge of the cheese and the cloth pulled down to the base. The cheese is then upended and the bottom surface and edge are also carefully scraped. The cloth is gently pulled away from the cheese, extra scraping being done where necessary.

Once the bandage has been removed, the cheese is placed on the smooth cloth and bathed in warm water at between 50–60°C (122–140°F). The cheese is immersed in the water for one minute and then removed and left to drain. The purpose of bathing is to remove the surplus fat from the surface. The formation of a rind is also encouraged by subjecting the cheese surface to the hot water.

The cheese is now returned to the cleaned mould and put back to press at $100kN/m^2$ (20cwt) for a further 24 hours.

Day Three

Materials needed for this day: a bench, a knife, a bowl, melted lard and bandage material.

The mould and cloth are removed from the cheese as before, care being taken not to damage the cheese surface. This is especially important at this stage – the cheese is to be bandaged and any damage will be permanent, unless the cheese is put back to press in a fresh smooth cloth for an extra day.

The trimmed cheese is placed on a clean board and measured for its bandages. A cap large enough to come down the sides is cut for the top and bottom of the cheese, as is a strip to encircle the sides. This is cut exactly to the height of the cheese. The cheese can be bandaged again on the fourth day and, if this is to be done, cut both sets of bandages at the same time – this will save time on the following day.

Bandaging is carried out to protect the cheese during storage and it is therefore important to ensure that the bandage is the correct size and tightly attached. The bandages are applied to the cheese using melted lard, as this helps prevent the cheese from drying out during storage.

The caps are applied first. Using a piece of material dipped into the lard, the melted fat is spread over the top and down the sides of the cheese and the cap is placed in position. The cap is smoothed on to the top of the cheese from the centre outwards using the fingers of both hands – this removes any trapped air. The cap is then smoothed on to

Fig 62 The cheese is capped at each end with a gauze bandage using lard or food grade paste. It is important to remove all air pockets from under the bandage.

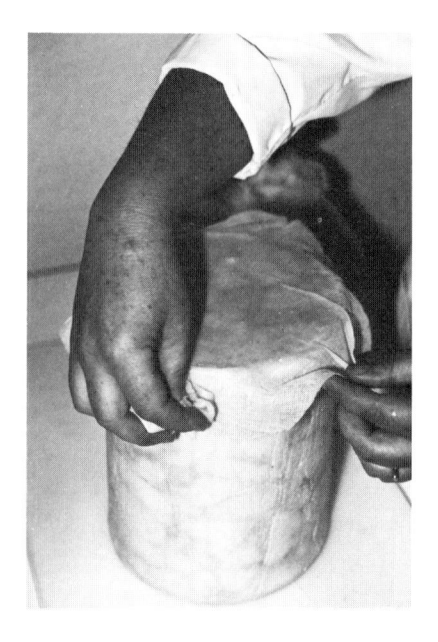

PRESSWORK AND STORAGE

Fig 63 The bandage around the cheese must be pulled tight to remove all air and to provide a little support to the curd.

Fig 64 The bandaged cheese is left to dry on a clean board.

the side of the cheese, the cloth being pleated as it is applied. It must be ensured that the lard is absorbed into the cloth, and extra lard can be used to help the material stick. The base of the cheese is capped in a similar way and the cheese is turned on to its side with the side facing the cheesemaker.

The upper surface of the side is well covered with lard, and the strip of bandage is positioned evenly across it and smoothed firmly into position. Working towards the cheesemaker the bandage is pulled and smoothed on to the cheese, keeping the edges even. The cheese is turned slightly and the procedure repeated until the strip completely covers the side of the cheese and overlaps itself by about 5cm (2in). The cheese is then returned to press.

Day Four

The cheeses are removed from the press and bandaged again. This bandage can be applied using lard or a polycellulose paste called *edifas*. The cheeses are then marked with their code, weight and date of manufacture and removed to the store. A suitable way to label cheese is to use a good quality tie-on label and to stitch this to the overlap bandage on the side of the cheese. Do not attach it to the base because it will become damaged or lost when the cheese is turned.

Storage

The cheese store must provide a cool, moist atmosphere that is clean and rodent-free. The store should be of adequate size and the cheeses should be stored on wooden or plastic-covered shelves which can be removed for cleaning. Temperatures for cheese storage range between 5–10°C (41–50°F) – cheeses stored above 12°C (54°F) ripen too quickly and those stored at below 5°C (4°F) ripen too slowly. Traditionally-prepared cheese which has been larded and bandaged, or brined, must be stored at a humidity of 80–90 per cent to prevent the cheese surface from cracking. Low humidities will cause the surface to crack and become contaminated with moulds and a humidity that is too high will lead to excessive mould growth and surface damage on the cheese. It is important to balance the humidity level so that it is high enough to prevent cracked rinds and low enough to keep mould growth to a minimum.

If the store is a larder or cellar, a bucket of water near the cheese will raise the humidity. In very hot weather a clean, damp tea-towel over the cheeses helps keep them cool and moist. Cheeses that have

PRESSWORK AND STORAGE

Fig 65   A larded cheese showing surface mould growth after three months storage.

Fig 66   A waxed cheese showing a date quite clearly.

been coated with wax or wrapped in film do not need a moist store.

After four weeks storage the traditional Cheddars develop a blue-green mould growth on their surface. This should be brushed off once a week to prevent it penetrating the cheese surface and causing damage. If a light brown dust suddenly appears on the surface of the more mature cheese this may be due to the cheese store pest *Tyroglyphus siro* (cheese mites). These mites live on the surface of the cheese and will cause damage to it. They can be kept in check by cleaning off all shelves as soon as they are empty and by making sure the cheeses are turned regularly. Diagnosis of mite infestation is simple. Watch the dust very intently – if it is formed by mites it will move (a magnifying glass may be of help). Prevention of mite infestation is the most effective method of control – if necessary, an infestation can be eliminated by using a formaldehyde fumigant in a sealed store that has been cleaned and emptied of cheese.

Cheese Faults

Cracked Rinds The rind of the cheese cracks and allows the penetration of mould during storage. This can be caused by draughts or by too low humidity in the store. Curd that is allowed to become chilled before moulding and does not knit well during pressing is prone to it, as is an over-acid cheese with a high moisture.

Discoloured Curd Curd may be streaky in colour if annatto is not added early enough or not well mixed in. Discolouration, combined with 'off' flavours and aromas, will be caused by contaminating bacteria in the cheese milk. Uneven cheddaring or salting of the curd will also cause slight discolouration. Bleaching may be seen in a very acid cheese.

Leaky Cheese This is cheese that leaks whey during storage. The problem may be due to a storage temperature that is too high, in combination with an over-acid cheese. Cooling the cheese may check the fault.

Blown Cheese The cheeses become 'blown' due to gas production by contaminating organisms such as coliforms and yeasts. Poor hygiene results in coliform contamination and causes small round gas holes throughout the cheese, accompanied by a 'cowey' taint. Yeast contamination will cause the whole cheese or the top of the cheese to blow up like a balloon, and it will have a sweet sickly taint. Yeasts,

PRESSWORK AND STORAGE

normally airborne, are sometimes able to enter the cheese during manufacture. These cheeses will have 'off' flavours and are best fed to stock.

Misshapen Cheese A cheese that is weak in body will not stand upright on the shelf, and the sides will bulge outwards. This is due to a lack of acid, probably caused by a slow starter or by a rushed process. The cheeses can be given support by wrapping a firm calico bandage round them, lacing it together at the front like a corset – this will prevent it from collapsing any further.

Cheese Ripening

The cheese is only half made when it reaches the store and is termed 'green'. During storage the curd texture and flavour change. The rate at which this occurs depends upon the variety of cheese, its moisture content and acidity, the starter culture used and the temperature and humidity of storage. For the average storage times for various cheese *see* Fig 67.

Chemical Changes During Maturing

The cheese constituents concerned with cheese ripening are the protein, casein, fat, lactose and water. During manufacture, the casein, fat and lactose are all changed in state by the enzymes found in the cheese.

Enzymes are biochemical catalysts which are able to alter the state of a substance without undergoing any change themselves – for example, the enzyme *lactase* will reduce lactose to glucose and galactose, which will then combine with water to form lactic acid. At

Cheese Ripening Periods

Caerphilly	10 – 21 days
Cheddar	6 – 12 months
Cheshire	4 – 9 months
Double Gloucester	4 – 6 months
Gouda	4 – 9 months
Leicester	4 – 6 months
White Wensleydale	2 – 3 weeks

Fig 67 The average maturing time for some semi-hard and hard pressed cheeses.

the end of the reaction the state of *lactase* remains unchanged.

Cheese ripening is not fully understood despite a great deal of research (most of it carried out at Reading University). The subject can be divided into two main areas – physical changes and chemical changes.

The combination of the two types of change produces flavour, body and texture in the cheese. These characteristics may take up to twelve months to develop fully in a hard cheese, and as little as three weeks in a semi-hard cheese.

Changes During Ripening

Physical Changes The physical state of the cheese is largely determined by the actions taken during manufacture. In the 'green' cheese the curd has a firm, resilient, rubbery texture and during storage this changes to a firm, waxy, velvety texture. The factors which control this development are the treatment of the curd during manufacture and the chemical changes which take place during storage. The initial firmness of the curd can be attributed to the moisture and fat content of the cheese. The springiness of the curd is determined by the scald temperatures used and the curd texture is partly due to the fat content of the cheese. Cheese with a high fat content has a smooth, velvety texture, and cheese with a low fat content has a coarse, short texture and tends to be crumbly.

Chemical Changes The chemical changes which take place concern the protein, fat and lactose. The protein changes are mainly due to the actions of the enzyme *rennin* which breaks down the casein in the cheese. This changes the body and texture of the curd and produces flavours. The protein is broken down into amino acids, many of which have considerable flavour properties which can be detected after eight weeks storage. The reduction of the protein changes the rubbery curd into a more mellow, waxy curd – this takes about four weeks.

The fat changes are brought about by the enzyme *lipase*. This enzyme is found occurring naturally in milk and is produced by many species of bacteria. The enzymes break down the fat into fatty acids and many of these produce distinctive flavours in cheese. The changes in the fat also influence the texture of the cheese, making it smooth and velvety. The lactose is broken down to form lactic acid which has two main functions – it provides an acidic environment for the proteolytic action of the rennin and it also contributes to the flavour of the cheese. The rate of the enzyme activity is dependent on the source of the milk, the type of starter culture, the method of cheese manufacture, and the storage conditions of the final cheese.

The changes which take place during ripening are complex, and it is the task of the cheesemaker to engineer all the manufacturing variables in order to produce a cheese curd which will mature into a good cheese, with properties characteristic of the particular variety. The skill required to make good cheese cannot be taught, it can only be learned by experience.

11
Grading and Packaging

Grading

Cheese cannot be manufactured to precise specifications and there will always be inconsistencies in milk quality and processing methods which, in turn, will produce cheeses of variable standards. It is, therefore, helpful to have a system of grading to assist in an assessment of the cheese – regular grading of cheese stocks will also reveal any problems. Grading is of particular importance when manufacturing cheese for sale because consistence in quality is the best marketing tool available.

The system of grading which follows is one that was set up by the Ministry of Food during wartime – after the war it was adopted by the various grading agencies and it has been in use ever since.

The cheese is graded at a set time, depending on variety. Cheddar, for example, is graded at four or eight weeks. The grading system considers three main qualities – flavour and aroma, body and texture, and colour and appearance. Points are allocated for each aspect, as below, and the total gives an indication of grade.

	Maximum Points
Flavour and Aroma	45
Body and Texture	40
Finish	10
Colour	5
	100

A top grade is 93 points, with not less than 41 for flavour and aroma, while a good grade is 85–92 points (not less than 38 for flavour and aroma).

Like cheesemaking, grading is an art acquired through long experience, but this does not mean that inexperienced grading has no value. It is always a good guide to cheese quality as long as it is carried out to a set of standards which are strictly adhered to. As experience is gained, grading will become more accurate.

Grading Method

Grading is carried out using a cheese iron or 'trier', a semicircular steel blade 11cm (5in) long and 1.75cm ($\frac{1}{2}$in) wide. The blade is attached to a 'T'-shaped handle.

The appearance of the cheese is noted – cracks or damage downgrade a cheese immediately.

The cheese iron is forced into the cheese and the resistance of the curd is assessed. The iron is turned one or two complete turns and the cheese plug is carefully withdrawn. As soon as the plug has been taken out, it is passed below the grader's nostrils to detect its aroma. To establish a clearer definition of aroma, a small portion of the cheese is removed from the plug in the iron. This is rubbed and rolled between the fingers until warm and the sense of smell is used to judge further aromas brought out by this warming – 'off' flavours can be detected easily at this stage. Points for flavour and aroma are then allotted.

Other assessments that are made are visual – the appearance of the plug of cheese will indicate the state of the fat, moisture and acidity of the curd. The plug is then gently replaced and the piece of warmed curd is used to seal the gap around it.

Assessment

Note the resistance of the curd as the iron is pushed into it. Is it weak, firm or resilient?

Judge the initial aroma of the cheese plug. Is it clean or has it 'off' flavours? And the aroma from the warmed curd. Is it 'clean' or has it 'off' flavours? An experienced grader can tell which grade a clean-flavoured cheese will develop into.

Assess the visual aspects of the plug. High acidity is indicated by a bleached appearance, and a cheese plug that does not fill the iron, leaving gaps between the cheese and the side, indicates a weak body. The fat content is shown by the smoothness of texture – free fat is undesirable. The moisture content is shown by the degree of resilience of the curd, and free moisture is also not a good sign.

The estimation of the body and texture depends upon a combination of several factors. Body is indicated by the initial curd resistance when the iron is forced into the cheese, and the effect of thumb pressure on the cheese surface and on the cheese plug. The curd

Fig 68 The cheese trier is sharply plunged into the cheese to cut the bandage and is then firmly pushed in and turned before withdrawing a sample of cheese.

should feel resilient but firm. A weak cheese has a 'soggy' feel to it due to its high moisture content. If this is suspected, lift the plug off the iron horizontally and see what happens – in a firm-bodied cheese the plug retains its shape while in a weak-bodied cheese it will bend and sag. A very hard cheese is usually an indication of either low fat content or a very high scald temperature.

Texture is indicated by the appearance of the cheese and by the state of the warmed curd. The smoothness of texture and the presence of cracks or gas holes is noted; the feel of the warmed curd then gives further indications of texture, and the descriptions used are 'smooth', 'pasty', 'flaky', 'crumbly', 'friable' and 'solid'.

Example

Cheddar at Four Weeks of Age The flavour and aroma of Cheddar at four weeks of age must be clean with no 'off' flavours. The body must be moderately firm and resilient and the texture must be smooth and flaky. The colour should be white, bright and uniform with a neat, regular, intact and smooth finish. Free moisture and gas holes must not be present.

Packaging

Packaging the cheese ready for sale requires a little market research. The packaging must please the eye of the purchaser, and it must also reflect the quality of the cheese.

The three main materials used for packaging are:

1. Heat-sealed vacuum packs.
2. Clear thin plastic film with cling properties.
3. A less malleable plastic (polypropylene) known as cellophane.

The decision as to which type to use must be based on financial limitations and market requirements.

Vacuum packaging will give relatively long-term storage and can be used for whole cheeses of up to 5kg (11lb), or for cut cheese. The system uses a plastic bag in which a vacuum is formed around the cheese before the bag is heat-sealed. Vacuum packaging gives protection to the cheese and prevents mould growth. It is not acceptable to all customers, many of whom prefer cheese in its 'natural state', and the equipment requires considerable initial capital outlay.

'Cling film' and cellophane will give good short-term storage and do not need any capital investment in machinery. The films will not give complete protection against mould growth and care must be taken during packaging and storage to prevent damage to the film.

Greaseproof paper can be used to wrap cheese which is cut at the time of purchase.

Small cheeses of 0.5–1kg (1–2lb) may be waxed for sale as a whole cheese. This is carried out by immersing a dry, mould-free cheese in food grade paraffin wax at 138–140°C (278–284°F) for 5–15 seconds. Waxing is an efficient way to coat cheese and can make an attractive package. The equipment required needs capital investment, but it is useful to remember that waxed cheeses do not need turning in the store.

Quality Control

Sales cannot be made until the product is of consistently good quality – to help ensure this, checks must be made throughout manufacture.

Raw Materials These include the milk and reagents which must be produced and stored efficiently. Milk is the main raw material and must never be used if its quality is in doubt. The use of a commercial milk testing laboratory may be helpful if any problems persist.

Process Control Equipment must be checked over regularly and the sheets for each cheesemake examined daily to check for any abnormalities during the process. Grading must be linked to the cheese make and to the starter record. Hygiene must be kept to a high standard and checked at intervals.

Product Inspection Random samples should be collected and stored at a range of temperatures similar to those that will be found in a customer's home. After a period of time the samples can be examined to detect any faults or 'off' flavours which may have developed.

Sales

When the combination of experience and quality control ensures consistently good cheese, sales can be confidently considered.

The legal implications must be checked first, as there is a considerable amount of legislation governing food production. The Cheese

Regulations (Food and Drugs Act), the Weights and Measures Act, Food Labelling Regulations (1980), and the Colouring Matter in Foods Act (1983) must all be considered, as well as the Environmental Health implications. The aim of all these regulations is consumer protection and it is advisable to go to the Local Authority Offices to gain advice before any sales are made.

Selling can be done at the farm gate, at a market, or in cheese shops and delicatessens. The method of selling and style of presentation will depend on the customer requirements.

The cheesemaker must educate the selling agency and the customer. He must make sure that they understand that cheese is a living product and that there are certain requirements which must be met to ensure the maintenance of its quality. Labels or advertising material can be informative, something that customers usually appreciate.

Showing

Many agricultural shows and food fairs are held and most of them have a dairy section where the cheesemaker can show his cheese. This can provide an informed assessment of the cheese and can also be helpful to advertise the product. Competition is keen and winning unlikely, but the experience is always useful.

It is customary to show the cheeses off to their best advantage and the traditional preparation of a cheese for competition can be carried out as follows.

The cheese is selected for flavour and body, ensuring that the surface is undamaged. The bandages are stripped and the cheese is scrubbed with warm water to remove surface mould. When dry, rough patches and blemishes are lightly rubbed with coarse sandpaper until they are clean, and when the surface is smooth and of even colour it is rubbed with olive oil. The cheese is then bandaged on the sides using a white lint bandage which leaves a 2cm ($\frac{3}{4}$in) space at the top and bottom. The bandage is attached by neat stitching at the join.

It is possible to have advertising material on the cheese after judging has taken place, and this may find potential buyers. The judges' grading is normally made available upon request, and this should be used by the cheesemaker as a comparative guide to how good his own grading is. A particular benefit of showing cheese is the opportunity it affords of meeting fellow cheesemakers, with the chance to exchange experiences and opinions.

12
Cheese Recipes

Caerphilly Cheese

Caerphilly cheese should be neat in appearance and uniform in size, weighing about 3.5kg (7.5 – 8lb). It should be evenly coated with white mould and when bored it should be white and firm, with a clean flavour and short texture.

Starter Three to four per cent inoculum for half an hour ripening or half to one per cent inoculum for two to two and a half hours. Temperature at starting 30°C (86°F).

Rennet Add rennet when acidity is about 0.19–0.2 per cent LA, and the temperature 30–31°C (86–88°F). Add at the rate of 30ml per 100 litres (1fl/oz to 20 gallons).

Cutting Lengthways and crossways with vertical knives and lengthways with horizontal knives, until curd pieces are the size of field beans. Acidity 0.13–0.14 per cent LA.

Scalding Stir gently without heating for 15 minutes, then raise the temperature to 33–34.5°C (92–94°F) in the next 15 minutes; stir for a further 15 minutes or until curd is fit. Fitness is determined by breaking a piece of curd in two; each portion should retain its shape.

Pitching Settle the curd about 45 minutes after cutting. When acidity is 0.15–0.16 per cent LA, run the whey off.

Running the Whey Cut a channel down the middle of the curd and pile the curd up on either side to form semi-cones. Using a curd knife, keep trimming thin slices of curd from around the outer edge of the semi-cone and pile them on top. This continues until the acidity is about 0.2 per cent LA. The curd is now soft and silky in texture.

Cut the curd into finger-length pieces and pile at the end of the vat. When acidity reaches 0.3–0.35 per cent LA, cut into 2.5cm (1in) cubes and salt at the rate of 95g per 100 litres (3oz to 20 gallons) of milk.

Moulding and Pressing Line moulds with cheese-grey, pack in curd until curd and mould weigh 5.5–6kg (12.5–13lb). Place moulds on top of each other and pull cloths up before moulds are put to press. Press at screw pressure for one to one and a half hours.

Remove cheese from press, wring the cloths out in salt water and put cheese back to press $20kN/m^2$ (2–4cwt) for three hours. Remove again and rub salt into the surface of the cheese. Put back to press $20kN/m^2$ (2–4cwt) until morning. Take cheese out and float in brine for 24 hours.

Brine An 18–20 per cent brine is prepared by dissolving salt in boiling water and allowing to cool. Solution should float an egg.

Ripening Dry cheese off and put on clean, dry boards until dry. Turn daily. Ripe in seven to twenty-eight days – white mould should be forming on the surface at seven days. Before marketing, rub the surface of the cheese with a mixture of rice flour and barley meal to give a uniform finish.

Cheddar Cheese

Milk Raw milk, cooled immediately after production, or milk pasteurised at 71.1°C (160°F) for 15 seconds. Initial milk acidity 0.14 per cent LA.

Starter Mixed culture of lactic organisms inoculated at temperature of 24–30°C (75–86°F) at a rate of one and a half to two per cent.

Ripening Incubated for half to one hour or until acidity rise of 0.04 per cent LA.

Renneting Milk temperature 30°C (86°F), rate of usage 25–30ml per 90 litres (1fl/oz to 20 gallons). Dilute the rennet in five or six times cold water. Add rennet throughout the vat and stir for four minutes. 'Top stir' for a further one to two minutes and leave the milk for 45–60 minutes.

Cutting When the curd is firm enough to cut (when it splits cleanly), it is cut lengthways and crossways with vertical and horizontal knives,

CHEESE RECIPES

and the acidity of the whey is tested. Cutting is continued until the desired curd particle size is obtained – wheat grain to small pea size. Acidity drop of 0.04–0.05 per cent LA.

Scalding The curds and whey are stirred gently as the temperature is raised about 1°C (2°F) every five minutes, increasing to 1°C (2°F) every three minutes, to a scald temperature of 39–40°C (102–104°F). Scald is one hour after cutting. Acidity at scald increases by 0.02–0.03 per cent LA.

Stirring After scald, the curds and whey are stirred for a further 45–60 minutes.

Pitching When fit, the curds are pitched. Acidity increases by 0.04–0.05 per cent LA to 0.2–0.24 per cent LA.

Whey Off The whey is drained off and the curd channelled. Acidity 0.24–0.28 per cent LA.

Cheddaring The curd is cut and blocked, then turned and piled every 10–15 minutes until the acidity reaches 0.75–0.85 per cent LA, a rise of 0.05 per cent every 15 minutes.

Milling Mill the curd at 0.7–0.85 per cent LA.

Salting Curd salted at a rate of two per cent salt, well mixed in, and the curds left to cool before moulding at 25.5°C (78°F).

Moulding Salted curd transferred to lined moulds.

Pressing Pressure is applied, gradually at first, and then increased to 75kN/m^2 (15cwt) for 12–16 hours; kept at 100kN/m^2 (20cwt) for the remaining days.

Presswork Routine – Traditional
Day 1 Put to press in rough cloth and increase the pressure gradually up to 100kN/m^2 (20cwt).
Day 2 Remove from press, carefully remove cloth, bathe in hot water at 60°C (140°F) for 30 seconds. Return to press in a clean, dry, smooth cloth. Repress.
Day 3 Remove from press, lard and bandage cheese, repress.
Day 4 Remove from press, lard and bandage and leave to dry on clean board. Label with number and code; put in cheese store.

Maturing Cheese store humidity 80–90 per cent; temperature 10–13°C (50–55.5°F). Turn cheese daily for one month, and weekly thereafter until mature.

Gouda (modified)

Hard pressed cheese with an open mealy body like Edam. It has a mild nutty flavour.

Milk The milk is pasteurised at a minimum of 67.8°C (154°F) for 15 seconds, maximum 71°C (160°F) for 15 seconds. Cool to 26°C (78.8°F). Check milk acidity – 0.14–0.18 per cent LA.

Starter A mixture of *Streptococcus lactic*, or *Strep. cremoris*, is added at a rate of one and a half per cent and ripened for 30–40 minutes, to give a rise of 0.03–0.04 per cent LA.

Rennet Added at a rate of 28ml/100 litres (just under 1fl/oz/20 gallons) milk. Milk temperature 30°C (86°F).

Cutting The curd should be firm enough in approximately 45 minutes. Cut the curd to wheat grain size. Acidity about 0.14 per cent LA.

Heating After cutting, allow the curd to settle, remove 15–20 per cent of the whey; replace with the same volume of water over a period of 15 minutes (water temperature 55°C or 131°F). Stir the curds and whey continually during the addition of the water to ensure even heating. Acidity after water addition 0.1 per cent LA. A temperature of 38°C (100°F) should be reached at the end of the 15 minutes. If not, make use of the water/steam jacket of the vat.

Acidity adjustment at this stage. If the acidity at cutting is higher than desirable remove a little more whey than normal and replace with more water. If the acidity is rather low at this stage remove less whey and replace with less water. Take care in both instances to adjust the temperature of the water to achieve a scald temperature of 38°C (100°F).

Whey Off The curd will be less firm than Cheddar. Draw back the curd and drain all the whey off.

Salting and Moulding The curd is salted immediately after whey off at a rate of seven per cent by weight. (1kg of cheese from 10 litres of milk.) Add the salt and mix well. Shovel the curd on to scalding cheese-grey cloth and tie in a Stilton bundle. Place the bundle on a rack and put another rack and a 25kg (56lb) weight on top of it. This racking and weighing helps to produce a close texture and encourages matting.

After 15 to 20 minutes the curd can be cut into shape and put into a lined mould.

Pressing Apply half an hour to one hour's light pressure (pull up the cloths) then medium pressure overnight ($75kN/m^2$ or 15cwt), increasing to $100kN/m^2$ (20cwt).

Ripening Time Two to six months.

Cheshire Cheese

Milk The milk is pasteurised at a minimum of 68°C (154°F) for 15 seconds, maximum 71°C (159.8°F) for 15 seconds. Cooled to 26°C (78.8°F). Check milk acidity (0.14–0.18 per cent LA).

Starter One to three per cent starter may be added. Normally two per cent with a ripening time of 30 minutes. Acidity rise during this period of 0.06 per cent LA.

Annatto Annatto is added at least 15 minutes before renneting to ensure even mixing. Additional rate 0.16–0.2 per cent, according to the degree of colour required.

Rennet Milk warmed to 30°C (86°F). Rennet is added at a rate of 32ml/100 litres (just over 1fl/oz to 20 gallons) milk. Dilute with cold water. Stir into the milk for four minutes.

Cutting After 40–60 minutes the curd will break evenly. Cut both ways with the vertical knife and both ways with the horizontal knife until about the size of baked beans. Acidity 0.13–0.15 per cent LA.

Scalding The curd is stirred continuously as the temperature is gradually raised. Stir for 15 minutes without heat addition. Stir and raise the temperature 0.5°C (1°F) every five minutes to 33–35°C (91.4–95°F) in 30 minutes. Times and temperatures adjusted according to the rate of acid production. Time from cutting to scald is 45 minutes.

Continue stirring until the curd is firm and free. Acidity approximately 0.2 per cent LA.

Drawing the Whey When the acidity is 0.21–0.23 per cent LA remove the whey. If the acidity rise is particularly rapid at whey off, dry stir the curd to encourage rapid drainage.

Texturing Cut the curd into blocks 20cm (8in) wide. This is the only time the curd is cut; afterwards the curd is broken by hand to keep a free, open texture in the cheese. Acidity at blocking 0.25–0.3 per cent LA. The curd blocks are broken in half and turned every 0.05 per cent LA rise, about every 10, 15 or 20 minutes. Acidity of 0.65 per cent LA – the curd blocks will be very small and the cheese ready to mill. (Block cheese milled at 0.5–0.55 per cent LA.)

Milling Four to four and a half hours from rennetting. Mill through a fine mill (twice through a cheddar mill). Keep well stirred.

Salting Dry salt at two per cent by weight; mix in well to ensure even ripening.

Moulding Weigh into a block mould and press at 49.6kN/m^2 (10cwt) for 24 hours.

Pressing
Day 1 Place moulds in oven or warm room at 21°C (69.8°F) overnight with *no* pressure.
Day 2 24.9kN/m^2 (5cwt) for two hours. Increase to 49.6kN/m^2 (10cwt) for the rest of the day.
Day 3 49.6kN/m^2 (10cwt).
Day 4 Pack.
Traditional cheese will be bandaged on Day 3 and Day 4.

Ripen in Store Two to six months for block cheese. Six to nine months for traditional cheese.

Lactic Cheese

(Made from surplus milk or starter.)

Milk Preferably pasteurised at 68°C (154°F) and cooled to 32°C (89.6°F).

Starter Approximately one per cent (enough to coagulate the milk overnight at 21–27°C or 70–80.6°F).

Method Hang up the coagulum the following morning in a thickly woven cloth – huckaback is particularly suitable – leave to drain in a warm room for 24 hours. Scrape down the sides of the cloth with a knife and mix. Hang up for a further 24 hours. By this time the cheese should be fairly firm in consistency and ready to salt.

Salt to taste and use right away. Lactic cheese will keep in a cold store for periods of up to one week. It may be mixed with sugar instead of salt, which makes it a suitable base for sweets. If made from surplus starter then hang up right away without leaving it overnight. Drainage may be hastened if the curdled milk or starter is heated to 49°C (120°F) for a few minutes before hanging up.

Cheddar Type of Smallholder Cheese

Take 30 litres (6.5 gallons) of milk (morning and evening mixed), add 300ml (10fl/oz) of starter and allow to ripen for one hour at 32°C (89.6°F). Stir in 4ml (0.15fl/oz) of cheesemaking rennet, stir for four minutes and leave covered over until it is firm (40–50 minutes). Cut into pieces the size of peas and, after stirring for 20 minutes, take out some whey, heat it to 49°C (120°F) and return it to the vat, thus raising the overall temperature. Repeat this until the heat is 35°C (95°F) and stir until firm (probably one hour after cutting).

Leave in the whey for half an hour; push the curd into a compact mass and run off the whey. Cut the curd into 7.5cm (3in) cubes, tie in a cloth and put under a board with a weight of about 18kg (40lb) on top – a stone could serve for this.

Open up the bundle of curd, break the pieces apart and tie up again every 20 minutes, until the curd smells slightly acid and feels rubbery – this will probably be about one and a half hours after running off the whey. Break the curd into pieces the size of walnuts and add 28g (1oz) salt for each 15 litres (3¼ gallons) of milk used.

Take a tin which is perforated to allow the whey to escape, and line with cheese-cloth; pack in the curd and put a board and weight on top to press the curd. Turn night and morning for two days; at the last turning put on caps and bands so that the cloth is pressed into the cheese. Keep in a well-ventilated place and turn every day for four weeks. The cheese will have a mild taste at this age – if a fuller flavour is wanted, keep for a few weeks longer.

CHEESE RECIPES

White Wensleydale

White cheese from 1.5kg (3lb) to 5kg (11lb), similar to Cheshire cheese.

Starter 0.2–0.4 per cent mixed culture starter added to milk at 21.1°C (70°F).

Ripening Temperature 30°C (86°F). Ripen for 45 minutes, or until acidity rises 0.01 per cent.

Rennet Temperature 29–30°C (84–86°F) according to ambient temperature. Quantity 30ml per 100 litres – diluted 6–8 times with cold water.

Cutting Curd firm with clean break in 35–45 minutes. Cut both ways with a vertical knife, 12.5mm (0.5in) blades, then cut with horizontal knife to size of large beans; acidity 0.13–0.245 per cent LA.

Stirring Stir to float curd for 5–10 minutes, settle in whey for 15–20 minutes and then refloat curd to scald.

Scald Scald 1.1°C (2°F) above renneting temperature and stir for 20 minutes.

Pitching Pitch until acidity is 0.135–0.155 per cent LA if curd is free, but stir to float if curd mats together.

Running the Whey Run off the whey when acidity is from 0.14 to 0.15 per cent LA. Ladle the curd on to a coarse (hessian) cloth or rack in drainer. Tie in a bundle. Open, cut and disturb, then retie in a bundle.

Texturing Disturb the curd in the bundles every 10 minutes until the acidity is 0.15–0.16 per cent LA. Then cut blocks and turn over. Repeat the process until acidity is 0.39–0.45 per cent LA.

Milling Cut or broken by hand or milled with a coarse peg mill.

Salting 1.6 per cent salt (1.8 per cent for wet cheese). Mix in thoroughly.

Moulding Mould small pieces tightly into the corners of the mould. Fill with remainder of curd loosely, but place small pieces on the top and corners. Leave overnight at 21.1°C (70°F).

Pressing Turn cheese into coarse hessian cloth and press for two hours at 24.9kN/m^2 (5cwt per cheese). Bandage with sewn-on bandages (gather top and bottom, stitch at intervals down side). Re-press for three hours. Dry off at 18.3°C (65°F).

Storing Store at 12.8°C (55°F) in a humid atmosphere for two to three weeks. Turn regularly in store.

Cheddar Cheese (DVI Starter)

Milk Raw milk, cooled, or milk pasteurised at 71.1°C (160°F) for 15 seconds.

Starter DVI culture equivalent to two per cent starter at 30°C (86°F).

Ripening 60–75 minutes. No visible rise in acidity.

Renneting Milk temperature 32°C (89.6°F), rate of usage 25–30ml/90 litres (1fl/oz to 20 gallons). Dilute rennet 5–6 times with cold water. Add rennet throughout vat. Stir for four minutes, then 'top stir' for two minutes. Leave milk to set for further 60–75 minutes.

Cutting Cut when curd is firm enough (softer than when using liquid starter). Cut lengthways and crossways with vertical and horizontal knives until pieces are pea-sized. Acidity drop of 0.05 per cent LA.

Scalding The curds and whey are stirred, and the temperature raised gradually to 40°C (104°F) in one hour. Acidity increase 0.02 per cent LA.

Stirring After scald temperature is reached, the curds and whey are stirred for a further 45–60 minutes.

Pitching When fit, the curds are pitched. Acidity increase 0.05 per cent LA.

Whey Off The whey is drained off the curd; curd is channelled. Acidity 0.2 per cent LA.

Cheddaring The curd is cut and blocked, then turned and piled every 10 minutes (each rise of 0.05 per cent LA) until acidity of 0.65–0.7 per cent LA is reached.

Milling Mill the curd at 0.65–0.7 per cent LA.

Salting Two per cent salt well mixed in and cooled to 25.5°C (78°F).

Moulding Salted curd transferred to lined moulds.

Pressing Pressure of 49.5kN/^{2} (10cwt), increased to 100kN/m^2 (20cwt) over two hours, for three days.

Presswork Routine
Day 1 Put to press in rough cloth.
Day 2 Remove from press, remove cloth, bathe in water at 60°C (140°F) for three minutes. Turn into a smooth cloth. Re-press.
Day 3 Remove from press, lard and bandage and leave. Re-press.
Day 4 Remove from press, lard and bandage, leave to dry on a clean board. Label with a number or code and remove to cheese store.

Maturing Turn cheese daily for one month and weekly thereafter, until the cheese is mature.

Double Gloucester

Double Gloucester is a large, round, flat cheese weighing 18–20kg (40–45lb). It is usually a medium 'red' colour and it has a clean, mild, mellow, slightly 'nutty' flavour. The texture is close with a firm body. Ripening time is four to eight months.

Milk The milk is pasteurised at a minimum of 67.7°C (153°F), maximum 71°C (160°F) for 15 seconds. Cooled to 26°C (79°F). Check acidity.

Starter A mixed culture of *Streptococcus lactis* or *Strep. cremoris* is added to the milk at the rate of two per cent.

Annatto Annatto is added at a rate of 15–30ml per 100 litres (½–1fl/oz per 20 gallons), at least 15 minutes before addition of rennet.

Ripening The milk is left to ripen for 30–45 minutes. Acidity rise of 0.07 per cent LA.

Renneting Temperature raised to 30°C (86°F). Rennet added 20–30ml per 100 litres (approximately 1fl/oz per 20 gallons). Dilute rennet 6–8 times its volume with cold water. Setting time 45–60 minutes.

Cutting When curd breaks cleanly, it is cut lengthways and crossways once with vertical and then horizontal knives, then crossways with a vertical knife until the pieces are the size of wheat grains. Acidity 0.14–0.16 per cent LA.

Scalding Stir the curd with great care for 15 minutes without any heat. Continue to stir and gradually increase the temperature to 37°C (98.6°F) in the next 30 minutes. Continue to stir until the acidity reaches 0.19 per cent LA. The curd should feel 'shotty'.

Whey Off The curd is pitched and the whey drawn off. Acidity 0.22 per cent LA. The curd is stirred to remove free whey and is then channelled and cut into blocks 20 × 30cm (8 × 12in).

Turning The blocks are turned and piled double after 20 minutes. This is repeated until the acidity is 0.55–0.65 per cent LA.

Milling The curd is milled once through a cheddar mill.

Salting The curd is well mixed and then salted at the rate of two per cent by weight.

Moulding The curd is filled into block moulds lined with disposable nylon cloths.

Pressing As for Cheddar – 100kN/m^2 (20cwt).

Storage Temperature-controlled store at 8°C (46°F) for four to eight months.

CHEESE RECIPES

Leicester

Red-coloured cheese with a clean acid flavour. It has a firm body and a moderately open, short, flaky texture. Four to six months ripening time.

Small Vat 180 litres (40 gallons)

Starter Milk temperature 27°C (80.6°F). Check milk acidity. Add one and a quarter per cent starter.

Ripening Time Milk temperature 27°C (80.6°F). Milk and starter left for half an hour; acidity increase about 0.02–0.03 per cent LA.

Annatto Add annatto (140ml/180 litres) 15 minutes before renneting. Adjust temperature to 30°C (86°F) ready for renneting.

Renneting At acidity of 0.17–0.18 per cent LA, add 55ml (2fl/oz) of rennet, diluted with six times its own volume of cold water, to the milk. Temperature 30°C (86°F). Immediately the rennet has been added, stir for four minutes, then leave undisturbed for about 45 minutes.

Cutting the Curd When the curd is fit (if it breaks cleanly when a finger is inserted and gently lifted to the surface), cut lengthways and crossways with vertical and horizontal knives, and then cut further using the vertical knife until fairly fine (wheat grain size). Note the acidity.

Scalding After cutting, the curd is stirred continuously. Temperature raised to 38°C (100°F) in 60 minutes. Acidity increased 0.15 per cent LA. Continue to stir for 20 minutes until firm and crumbly.

Pitching Whey acidity 0.14–0.15 per cent LA. Leave pitched for 20 minutes.

Whey Off Acidity taken from whey squeezed from curd – 0.2 per cent LA (time from addition of starter to whey off is two and a half hours).

Turning Channel curd down centre to facilitate drainage; cut into blocks 20cm (8in) wide, turn and move down vat. Turn blocks at five minute intervals (longer intervals if acid production is slow, shorter intervals if acid production is fast); as acidity increases, lengthen the turning interval slightly. Continue until acidity is 0.45 per cent LA.

Milling Mill once through cheddar mill.

Salting 20g/kg (1oz/3lb) curd.

Pressing As for Cheddar.
Blocks – 100kN/m^2 (20cwt) for two days, then pack.
Traditional – Day 1 Bathe at 80°C (176°F); smooth cloth, press.
              Day 2 Lard and re-press.
              Day 3 Edifas (*see* page 110), dry and store.

Appendix

Titratable Acidity of Milk

The most commonly-used method of determining the acidity of milk is by titration with a standard alkali. Lactic acid may be titrated with Sodium Hydroxide solution using Phenolphthalein as an indicator.

$$NaOH + CH_3CH(OH)COOH = CH_3CH(OH)COONa + H_2O$$

One molecular weight of Sodium Hydroxide, therefore, neutralises one molecular weight of lactic acid. Thus, 40g of NaOH neutralises 90g of lactic acid. As a normal solution of NaOH contains 40g of this substance per litre it follows that:

1 litre of N NaOH neutralises 90g of lactic acid
1 litre of N/9 NaOH neutralises 10g of lactic acid
1ml of N/9 NaOH neutralises 0.01g of lactic acid

Suppose that 10ml of milk are titrated with N/9 NaOH and that 3.9ml of this solution were required. If all the acidity is due to lactic acid, 3.9ml of N/9 NaOH neutralise 0.039g of lactic acid.

On the assumption that 10ml of milk would contain 0.039g, the percentage 0.39 is obtained by dividing the actual quantity of N/9 NaOH by 10.

Glossary

Acidity The condition of the milk, whey, or cheese curd at various stages of manufacture, expressed as a percentage of lactic acid present in the sample tested.
Annatto A harmless colouring matter extracted from the berries of the South American plant *Bixa orellana*, and used for colouring cheese.
Bacteriophage Bacterial virus which destroys the susceptible host (lactic streptococci).
Bandaging Covering the cheese with cloth for protection and to retain its shape.
Bathing The immersion of cheese in hot water in order to form a rind.
Body Degree of firmness in the substance of the cheese.
Brining The immersion of cheese in a solution of brine.
Caps Square or round pieces of calico or muslin which are placed over the top and bottom of some varieties of cheese in conjunction with bandages.
Cheddaring Treatment of the curd, following the removal of the whey from the vat, to bring it into a sufficiently dry, firm and acid condition for milling.
Cheese-grey Unbleached fabric used for covering certain varieties of cheese.
Coagulation (Clotting or setting.) The precipitation of the casein in milk, following renneting, forming the curd or coagulum.
Coagulum The soft, semi-solid state of the milk resulting from the action of rennet.
Cooler A large, shallow, rectangular pan into which all the curds from a vat are placed for continuation of the process after pitching.
Cracked Rinds Splits in the rind of the finished cheese.
Curd The solid mass resulting from the cutting and heating of the coagulum formed by the clotting of milk by rennet or acid.
DVI Direct vat inoculation starter culture which is in powder form and added directly to the fresh milk.
Denatured Change in state.

GLOSSARY

Edifas Polycellulose paste used for bandaging cheese.
Film Wrapping Plastic covering in which cheese is ripened or, alternatively, in which it is subsequently wrapped.
Gassiness Production of gas in the curd or the cheese resulting from bacterial action, causing gas holes or 'blowing'.
Grading The assessment of cheese quality for marketing.
Homogeneous Of uniform state.
Hot Iron Test A test to measure the degree of acidity developed in the curd as it matures.
Micelle A complex molecule of the protein casein.
Milling Passing the mature curd through a mill which tears or cuts it into irregular pieces.
Moulds Containers of various sizes and shapes, according to the variety of cheese, into which the milled curd is placed before pressing.
Moulding The packing of cheese curd into a prepared mould ready for pressing.
Piling the Curd The mass of curd in the vat, after the whey is run off, is cut into blocks, piled at the sides of the vat, leaving the centre of the vat clear to facilitate drainage. The piled curd is cut, turned and progressively piled again.
Pitching Settling of the curd to form a mass in the bottom of the vat after scalding, before the whey is run off.
Pressing The application of pressure (in amounts varying with different varieties of cheese) to the milled curd in the moulds.
Rennet A commercial preparation containing *rennin*.
Renneting The addition of rennet to milk to cause coagulation.
Rennin A protein-splitting enzyme, prepared from the fourth stomach of the calf, and possessing the property of clotting milk.
Rind The natural integument formed by the drying of the surface of the cheese.
Ripening of Cheese The storage of cheese after removal from the cheese moulds, under controlled conditions of temperature and humidity, varying with different varieties of cheese.
Ripening of Milk Development of lactic acid in the milk to be used for cheesemaking.
Running or Weeping Seepage of whey from cheese during storage.
Salting The addition of salt to the curd, usually carried out at the time of milling.
Scalding Heating the curds and whey in the vat after cutting.
Starters Cultures of lactic streptococci added to milk to promote acid development.
Sweating Loss of moisture from cheese during ripening, due to high temperature.

Texture A term indicating the visual openness or closeness of the cut surface of the cheese.
Vat A double-jacketed tank in which the milk to be used for cheesemaking is coagulated and subsequent processes are carried out.
Whey The watery part of milk which remains after the separation of the curd by coagulation.

Further Reading

A Brief Survey of Plant Coagulants, J. Burnett (United Trade Press) 1976

Cheese, Volume 1 – Basic Technology, J.G. Davis (J. & A. Churchill) 1965

Richmond's Dairy Chemistry, J.G. Davis and F.J. McDonald (Charles Griffin & Co Ltd) 1952

Cheese Starters, J.E. Lewis (Elsevier Applied Science Publishers, London and New York) 1987

The Textbook of Dairy Chemistry, Vols 1 and 11, E.R. Ling (Chapman & Hall Ltd) 1963

Milk and Dairy Regulations, Ministry of Agriculture (HMSO, London) 1959

Cheesemaking, Bulletin No. 43, Ministry of Agriculture HMSO, London) 1959

Cheese Regulations, Ministry of Agriculture (HMSO, London) 1970

Cheesemaking Practice, R. Scott (Applied Science Publishers Ltd, London) 1981

Pasteurizing Plant Manual, Society of Dairy Technology (72 Ermine Street, Huntingdon, Cambridgeshire) 1983

Useful Addresses

Suppliers of Starter Cultures

Liquid Cultures

Somerset College of Agriculture
 and Horticulture
Cannington
Bridgwater
Somerset

Dairy Cultures Ltd
The Park
Castle Cary
Somerset

Freeze-Dried Cultures

Eurozyme Ltd
13 Southwark Street
London SE1 1RG

Chr. Hansens Laboratory
476 Basingstoke Road
Reading
Berkshire

Miles Laboratories Ltd
PO Box 37
Stoke Poges
Slough
Buckinghamshire

Suppliers of Rennet

Aplin and Barrett Ltd
Trowbridge
Wiltshire

Marschell Division
Miles Laboratories Ltd
PO Box 37
Stoke Poges
Slough
Buckinghamshire

Chr. Hansens Laboratory
476 Basingstoke Road
Reading
Berkshire

Dairy Cultures Ltd
The Park
Castle Cary
Somerset

Suppliers of Equipment

Smallholding Supplies
Little Burcot
Nr Wells
Somerset

James Gilmore
Centri-Force Engineering Co.
 Ltd
73/79 Dykehead Street
Queenslie Industrial Estate
Glasgow

R. and G. Wheeler
Hoppins
Dunchideock
Exeter
Devon

R.J. Fullwood and Bland Ltd
Grange Road
Ellesmere
Shropshire

Suppliers of Laboratory Equipment

Smallholding Supplies
Little Burcot
Nr Wells
Somerset

Borolabs Ltd
Paice's Hill
Aldermaston
Berkshire

Index

Note: italic numerals denote page numbers of illustrations.

Acidimeter 22, 24, *54*
Acidity 56, *57*, 75, 90, 97, 99, 102
 test 53, *55*
Additives 40
Aluminium 13
Annatto 45, 64, 75, 98
Antibiotics 27, 33

Bacteria 13
Bacteriophage 32, 33, 50, 98
Bandaging 14
Burette 22

Calcium 27, 30, 40
Case hardening 101
Casein 27, 40
Cheese
 classification 10
 craft 7, 8, 9
 definition 10
 green 113
 storage 71, 110
 types 10
 utensils *23*
 yields 29, *63*, 65
Cleaning
 agents 50
 equipment 49
 methods 48, 50, *51*
Cloths 14, 91
Colouring 45
Curd 15, 18, 22
 brushes 15
 cheddaring 68, *69*, *70*, *87*, 88, *90*, *91*, 103
 cooking 65, 67, 81, 101
 cutting 65, *66*, 78, *79*, 80, *80*, 99, 100
 formation 40, 99, *100*
 milling 68, 91, 103

moulding 71, 91, *92*, *94*, 104
pitching 67, 82, *83*
pressing 71, 95, *95*, 104
renneting 40, *41*, *42*, *43*
salting *51*, 68, 91, 103
stirring 67, 82, *82*, 102

Effluent 52
English cheese 9
Equipment 13-24
Ewes' milk 9, 81

Fat 27
 globules 27, 28
Faults 33, 112
Flavours 26, 29, 31, 103
Followers 18
Freezing milk 26

Grading 117, *118*, 119

Health 26
Holder tank 30
Hot iron test 56, *58*, 98
Humidity 53, 59, *60*, 110
Hygiene 48-51
Hygrometer 60, *60*

Inhibitory substances 33

Jacket, vat 15

Lactase 113
Lactation 26, 44
Lactic acid 13, 29, 31
Lactose 27, 35, 114
Lipase 114

Mastitus 25, 27
Maturing 113-114

143

INDEX

Milk
 abnormal 26
 cans 14
 Channel Island 26, 27, 43, 97
 composition 27, *28*
 cooling 23
 cows' 27
 ewes' 28, 34, 44, 81
 frozen 96
 goats' 28, 34, 43
 hygiene 25
 plunger 18
 protein 28, 40, 56
 quality 25-27, 64
Milking 25
Mites 112
Mould growth 107, 110, *111*, 112

Packaging 119
Pasteurisation 29
Pathogenic bacteria 29
pH meter 56
Phenolphthalein 22
Press
 types of 19, *20*, 23, *24*
 mechanics of 21, *21*
Presswork 105-110, *106-109*

Quality control 120

Recipe
 contents 11
 function 10
 ingredients 11, 63
Recipes
 Caerphilly 122
 Cheddar 71, 123
 Cheddar DVI 130
 Cheshire 126
 Double Gloucester 131
 Gouda 125
 Lactic 127
 Leicester 133
 Smallholder 128
 White Wensleydale 129
Records 61, *61*
Regulations 25
Remedial actions 96-104
Rennet 65, 40, 98
 action *41*, 42

 microbial 44
 plant 44
 rates of use 42, 44, 76, *76*, 77
 storage 45
 vegetarian 44
Ripening
 initial 64, 75, 98
 during storage 103, 113, *113*
Romans 9

Sales 120
Salt
 function 46, 48
 quality 47
 rates of use 46
 storage 47
Shows 121
Smoking 47
Sodium hydroxide 22, 53
Sodium hypochlorite 33, 50, 51
Stainless steel 13
Starter 31, 64, 96, 97
 bacteria 31
 bulk 37
 direct vat inoculation 38, 42, 99
 failure 33, 96
 frozen 37
 milk 34
 propagation 34-38, *35*, *36*, *38*
 records 39
 slow, causes of 32, 96
 temperatures 32
Stilton bundle 84, *85*, *87*

Taint 13
Temperature 53, 58, 110
Texture 31
Thermometer 14, 57, *59*
Titratable acidity 135
Toxins 27

Vat
 design 13, 15, *16*, *17*
 kitchen 22
 purchase 16
Vitamin B$_{12}$ 50

Water supply 52
Waxed cheese *111*, 112
Whey 33, 52, 95, 101

144